Radiation and Health

Radiation and Health

Thormod Henriksen and
H. David Maillie

Taylor & Francis
Taylor & Francis Group
LONDON AND NEW YORK

This edition based on the Norwegian edition Stråling og Helse
(Radiation and Health) published 1993 by The Institute of Physics at
The University of Oslo.

English edition first published 2003
by Taylor & Francis
11 New Fetter Lane, London EC4P 4EE

Simultaneously published in the USA and Canada
by Taylor & Francis Inc,
29 West 35th Street, New York, NY 10001

Taylor & Francis is an imprint of the Taylor & Francis Group

© 2003 Taylor & Francis

Printed and bound in Great Britain by
Biddles Ltd, Guildford and King's Lynn

This book has been produced from camera ready copy supplied by
the authors

Every effort has been made to ensure that the advice and
information in this book is true and accurate at the time of going to
press. However, neither the publisher nor the authors can accept any
legal responsibility or liability for any errors or omissions that may
be made. In the case of drug administration, any medical procedure or
the use of technical equipment mentioned within this book, you are
strongly advised to consult the manufacturer's guidelines.

British Library Cataloguing in Publication Data
A catalogue record for this book is available from the British Library

Library of Congress Cataloging in Publication Data
A catalog record for this book has been requested

ISBN 0-415-27162-2 (pbk)
ISBN 0-415-27161-4 (hbk)

Table of Contents

Preface

Radioactivity and ionizing radiation have been known for more than 100 years. We are all living in an ocean of radiation from cradle to grave. A modern society uses radiation for a number of purposes such as cancer therapy, medical diagnostics, as well as for research and in industry. A significant fraction of the world's electrical energy production comes from nuclear energy, which results in the production of considerable amounts of radioactive fission products and plutonium. The latter is considered by some people to be a dangerous waste and by others as a valuable resource.

Through the last century the research community has gained a large amount of information on radiation physics and the effects of radiation on plants, animals and humans. The purpose of the present book is to present some of this knowledge to the public. Information about radiation and health is very important if radiation is to be used properly. We know that radiation can only yield a biological effect when given as a physical dose, i.e. when radiation energy is absorbed in a biological system. We have paid special attention to dosimetry and doses from environmental sources and radioactive isotopes. Therefore, we hope that readers would be able to carry out some rather rough dose-estimates and thus be able to judge for themselves the health risks involved. Some of the basic steps in radiobiology are given.

The present book is a translation and an update of a Norwegian book with the same title. The book has been used for a number of courses at the undergraduate level on environmental issues. The first printing of the book came shortly after the Chernobyl accident. The accident upset the whole world and also demonstrated that public knowledge of radiation physics and radiobiology was lacking. We, therefore, made efforts to present information on the nature of radiation and its effects on biological systems. The Norwegian book was updated both in 1993 and in 1995. The present book includes some of the very interesting research which has been carried out in recent years.

Oslo and Rochester, April 2000

Acknowledgements

A large number of people: colleagues, students and teachers have made this book possible. For the Norwegian book, a number of colleagues have contributed in a very positive way. They include my (T.H) very first teacher in radiation physics at the University of Oslo, Anders Storruste. Furthermore, my colleague Finn Ingebretsen, as well as my former student Tove Svendby, have made significant contributions. Thanks also go to Erling Stranden and Terje Strand from the Norwegian Radiation Protection Authority for their contributions, in particular on radon measurements. Also Per Wethe from Institute for Energy Technology, made a significant contribution in Chapter 13. We are also grateful for the cooperation of Arian van der Woude of KVI (Kernfysish Versnellere Instituut, Groningen, Netherlands). We are very thankful to them all for all discussions, research data and figures.

We have used a number of drawings and illustrations throughout the book. We would like to take this opportunity to thank Trine Sperstad Tveter and Per Einar Arnstad for these drawings.

The present English version of the book is a cooperation between University of Oslo and University of Rochester. This cooperation has been mediated through Professor Bill Bernhard, and we would like to thank him for significant contributions with editing the book as well as discussions on central radiation physics issues. In addition, Kermit Mercer, Michael Debije, and Audrey Debije assisted in editing the text.

Finally we would like to thank H. Bjercke and A. Westerlund at the Norwegian Radiation Protection Authority, A. Skretting at The Norwegian Radium Hospital and Rolf Falk at the Swedish Radiation Protection Institute for valuable data which are included in the book.

Chapter 1

Radiation is Discovered

Introduction

From the beginning of life on earth, all living things have been exposed to radiation. Life started and developed in spite of, or quite possibly because of, radiation. It is disquieting to people that they coexist with radiation yet it cannot be seen, heard or felt.

Radiation, when broadly defined, includes the entire spectrum of electromagnetic waves: radiowaves, microwaves, infrared, visible light, ultraviolet, and x-rays and particles. This book is concerned with radiation having energies high enough to ionize matter. Examples are x-rays, cosmic rays, and the emissions from radioactive elements. Although the term "ionizing radiation" is in this case more precise, common usage often omits "ionizing" and this is what is done here. In this book, "radiation" means "ionizing radiation".

Prior to the reactor accidents at Three Mile Island in the United States and at Chernobyl in the former Soviet Union, radiation issues were addressed primarily by specialists. Now, however, radiation and biological effects are debated by the public and political leaders. They use expressions such as: radiation dose, becquerel, sievert, cesium and γ-radiation. Because people are easily confused by this technical language, all too often they are left with the perception that *all* uses of radiation are dangerous.

This book is written for those who want to understand radiation in order to make informed decisions about it in their lives. This field of science, founded at the turn of the century, has provided dramatic insights into physics, chemistry, biology, and medicine. The work of the early investigators provided a strong foundation from which to understand radiation phenomena. We will meet a few of them in the following pages and gain insight into their work and lives.

Radioactivity

Since radiation can neither be seen nor felt, it is a challenge just to know whether it is present or not. The situation is the same for radio and TV signals; you can not see or feel them. But if you have a radio tuned to the correct frequency, it will detect the presence of the radio signal. Similarly, in order to detect radiation, a special detector or sensor is needed.

Henri Becquerel was using a photographic plate when he discovered radioactivity in the spring of 1896. Becquerel was 44 years old and was working with compounds that could emit light after being exposed to sunlight (called fluorescence). The light from the exposed samples was then detected by these photographic plates. Becquerel had a compound, uranium salt, in one of his desk drawers. He had no idea that uranium salt could emit radiation. During a period of cloudy weather in Paris, Becquerel was unable to carry out his usual experiments involving fluorescence induced by sunlight. Instead, he decided to check for any possible light leaks by developing some of his unexposed photographic plates. To his surprise, he found that a plate on top of the uranium salt was black. Somehow, intense radiation had exposed the plate. Becquerel was puzzled at first then realized that some unknown type of radiation had to be coming from the uranium salt – *radioactivity was discovered.*

Marie Curie and her husband Pierre Curie worked to isolate these radioactive materials from the parent rocks. After a large amount of work, they isolated two radioactive elements. The first one was called polonium (after Marie's homeland Poland), and the other one was called radium (which is "the thing that radiates"). Marie Curie died in 1934 from a blood disease, possibly leukemia, which may have been caused by her work. She was 67 years old.

X-rays

X-rays have much in common with the radiation from radioactive compounds. This type of radiation was discovered by Wilhelm Conrad Roentgen at the University of Würtzburg in Germany. He, like many others in those days, was studying electric discharges in glass tubes filled with various gases at very low pressures. In experiments on November 8, 1895, Roentgen had covered the tube with some black paper and had darkened the room. He then discovered that a piece of paper painted with a fluorescent dye would glow when he turned

The Curie Family

© The Nobel Foundation

Curie is a prominent name in radiation science. For their work, the Curies won 3 Nobel prizes. Marie and Pierre shared the prize in physics with Becquerel in 1903. Marie got the prize in chemistry in 1911, and finally, their daugther Irene (the girl in the picture below) won the prize in chemistry in 1935 together with her husband Frederic Joliot.

Marie Curie (1867–1934) came from Poland and her name was Sklodowska before she married Pierre Curie in 1895. Marie was very gifted and worked all her life with radioactive compounds. She discovered the elements radium and polonium, the latter being named after her homeland.

It was the purification of radium that earned her the Nobel prize in chemistry. She is the only person who has ever won the Nobel prize in both physics and chemistry.

UK Science Museum/Science & Society Picture Library

Pierre Curie (1859–1906) worked with radioactive compounds with Marie, but he is also well known for his work in magnetism. Named after him we have "the Curie point", "Curie's law" and "the Curie constant".

Pierre was only 47 years old when he died in a traffic accident (involving a horse drawn carriage) in Paris.

Irene Joliot Curie (1897–1956) was the oldest of the two girls in the Curie family. She became a physicist and married **Frederic Joliot** (1900–1958). Together they discovered man-made radioactivity in 1934. By bombarding aluminum with α-particles, they produced a radioactive isotope of phosphorus, P-30.

The two giants

who discoved radioactivity and x-rays

© The Nobel Foundation

H. Becquerel

Henri Becquerel (1852–1908)

Becquerel was French, a third generation professor of physics. On March 2nd of 1896 he discovered radioactivity. For this discovery, he was awarded the Nobel prize, together with the Curies, in physics in 1903.

The unit for the intensity of a radioactive source is named after Becquerel. Thus, 1 becquerel (abbreviated Bq) indicates that, on average, one atom in the source disintegrates per second.

W. C. Roentgen (1845–1923)

Roentgen discovered x-rays in the fall of 1895. He immediately understood that the radiation from the x-ray tube has special properties, for example, it was possible to "see into" a human body. Within months this new radiation, called x-rays, was used in medical diagnostics.

It was realized that x-rays could also kill living cells, and that the sensitivity for killing varied from one cell type to another. It was easier to kill cancer cells than normal cells. Consequently, it could be used in cancer therapy.

The unit R (roentgen) used for radiation exposure was named after him. An exposure of 1 R means that the radiation dose to ordinary tissue is approximately 9.3 mGy (see Chapter 4).

© The Nobel Foundation

W.C. Roentgen

on the high voltage between the electrodes in the tube. Realizing the importance of his discovery, Roentgen focused all his attention on the study of this new radiation that had the unusual property of passing through black paper. He found that the radiation not only could penetrate black paper but also thick blocks of wood, books and even his hand. In the dark room, he viewed shadows of the bones in his own hand. This was the first x-ray image. The German anatomist von Koelliker proposed that the new type of radiation be called *Roentgen rays*. Although this term is used in many countries, the most common name used is that coined by Roentgen himself, x-rays. The letter "x" is often used by physicists to indicate an "unknown". Since the nature of these rays was unknown, Roentgen called them x-rays.

It is interesting to note that, more than 100 years ago, the two great discoveries of radioactivity and x-rays occurred within 4 months of each other. Most x-rays are from man-made sources. While x-rays do emanate from natural sources in outer space, they are absorbed by the upper atmosphere and do not reach the Earth's surface. Radioactivity, however, comes from both man-made sources and natural sources which are on the surface and deep within the Earth.

Ionizing Radiation

The common name for both radiation from x-ray machines and radioactive sources is *ionizing radiation*. The name indicates that the radiation has sufficient energy to ionize atoms and molecules. An ionization takes place when an electron is removed from its position in the atom or molecule. Since a molecule usually has no net charge to begin with, the loss of a negative electron leaves behind a positive ion. The electron can then end up on another molecule which then becomes a negative ion. The creation of positive and negative ions in matter is the signature of radiation, allowing us to detect and categorize it. Ionizing radiation is distinct from low energy radiations which include ultraviolet, visible, infrared, microwaves, and radio waves that produce effects, for the most part, of a different nature.

The mysterious x-rays

In the first period after the discovery of x-rays many people had a number of strange ideas as to what x-rays really were, and how they could be used. Here are a couple of examples of what one could see in the newspapers and magazines at that time.

THE NEW ROENTGEN PHOTOGRAPHY.
" LOOK PLEASANT, PLEASE."

The newspapers very often had headlines such as *Electric Photography Through Solid Bodies* and *Photography of Unseen Substances*. The drawing to the left is from *Life* magazine, February 27, 1896. Here we see a common misunderstanding. Some people believed that it was possible to take x-ray pictures with reflected x-rays. This means that the x-ray tube and film is in the photographer's box, as shown.

(From R.F. Mould (1993), reproduced with permission from R.F. Mould)

X-rays can easily penetrate a living body, providing us with a powerful tool in medical diagnosis. In the first period after the discovery, people had a lot of ideas of what you could see, for example, through clothes. Some rumours said it was possible to watch people when they changed into swimming suits inside small cabins on the beach. It is, therefore, not so surprising that a London tailor company advertised that they could make *x-ray proof underclothing for ladies*. The drawing to the right was used as an advertisement for x-ray proof underwear.

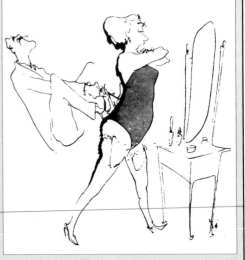

(From R.F. Mould (1993), reproduced with permission from R.F. Mould)

Chapter 2

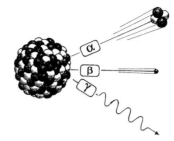

What is Radioactivity?

Radioactive Elements

The atomic structure of most elements contains a nucleus that is stable. Under normal conditions, these elements remain unchanged indefinitely. They are *not* radioactive. Radioactive elements, in contrast, contain a nucleus that is unstable. The unstable nucleus is actually in an excited state that can not be sustained indefinitely; it must relax, or *decay*, to a more stable configuration. Decay occurs spontaneously and transforms the nucleus from a high energy configuration to one that is lower in energy. This can only happen if the nucleus gives off energy. The energy emitted by the relaxing nucleus is radiation. All radioactive elements have unstable nuclei; that is what makes them radioactive.

The Nature of Radiation

The energy emitted by an unstable nucleus comes packaged in very specific forms. In the years that followed the discovery of radioactivity, determining the kind of radiation emitted from radioactive compounds was of great interest. It was found that these radiations consisted of three types called: alpha (α), beta (β) and gamma (γ) radiations after the first three letters in the Greek alphabet (see Figure 2.1).

The nuclear emission transforms the element into either a new element or a different isotope (see page 10) of the same element. A given radioactive nucleus does this just once. The process is called a decay or a *disintegration*.

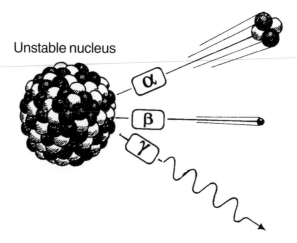

Unstable nucleus

Figure 2.1. A radioactive atom possesses an unstable nucleus. This means that radioactive atoms will emit radiation sooner or later and convert into a more stable state. The types of radiation that may be emitted are called alpha (α), beta (β) and gamma (γ) radiation.

The evidence for the three types of radiation comes from an experiment in which the radiation from radioactive compounds was passed through a magnetic field. γ-rays passed through the field without disturbance, whereas the two other types were deflected from a straight line. Because it was known at that time that charged particles are deflected when they pass through a magnetic field, the conclusion was evident; γ-rays have no charge while α- and β-radiations consist of charged particles. The α-particles, deflected in one direction are positive whereas the β-particles, deflected in the opposite direction, are negative.

Alpha radiation

In 1903, Ernest Rutherford (a New Zealander who worked in Cambridge, England most of his life) performed a simple and elegant experiment showing that the α-particle is the nucleus of the helium atom. Rutherford positioned one glass tube inside a second glass tube. The inner tube contained a radioactive source that emitted α-particles (see Figure 2.2). The outer tube contained a vacuum and at each end there was an electrode. The α-particles passed through a thin window, picking up two electrons on the way, and entered the outer tube as a gas. When Rutherford turned on the high voltage between the electrodes, the tube emitted light at very specific wavelengths (specific colors). He compared wavelengths of this light with the wavelengths of light produced by a similar tube that he had filled with helium gas. The colors of the light were identical. Rutherford concluded that an α-particle is simply the nucleus of a helium atom and that when the α-particles reach the outermost tube they have picked up two electrons to become helium atoms.

Ernest Rutherford

AIP Emilio Segrè Archives, courtesy Otto Hahn and Laurence Badash

Figure 2.2. A drawing of Rutherford's experiment. α-particles from a radioactive source pass through a thin window into an evacuated glass tube. The α-particles "pick up" electrons and become ordinary helium atoms. The color of the emitted light confirmed that the glass tube became filled with helium.

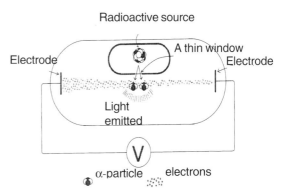

Beta and gamma radiation

Experiments have shown that the β-particle is a fast moving electron, whereas γ-radiation is an electromagnetic wave. Other examples of electromagnetic radiation are ultraviolet (UV), visible light, infrared and radio waves. Electromagnetic radiation is characterized by its wavelength or frequency. The wavelength is the distance from one wave peak to the next and the frequency is the number of waves passing a given point per second. Through quantum mechanics it is known that particles can be described as waves and vice versa. Thus, γ-rays and other electromagnetic radiation are sometimes described as particles and are called *photons*.

What is an isotope?

In several places throughout this book isotopes are mentioned. Some isotopes are unstable, and therefore radioactive, while others are stable, and thus non-radioactive. What is an isotope? An element can exist in several versions that are chemically equivalent but have different atomic weights. The atomic weight of an element can be changed by altering the number of neutrons in the nucleus. This is illustrated below for the most common of the elements, hydrogen.

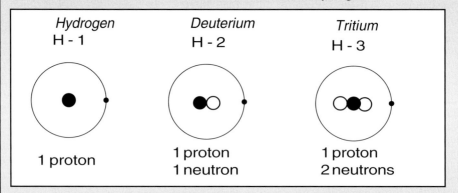

| Hydrogen | Deuterium | Tritium |
| H - 1 | H - 2 | H - 3 |

| 1 proton | 1 proton | 1 proton |
| | 1 neutron | 2 neutrons |

The nucleus of an atom consists of protons and neutrons (called nucleons). The number of protons determines the element and the number of nucleons determines the atomic weight. Isotopes are atoms with the same number of protons, but with different numbers of neutrons.

Isotopes are written using the symbol for the element, such as H for hydrogen, O for oxygen, and U for uranium. Furthermore, the nucleon number is used to separate the isotopes. For example the three hydrogen isotopes mentioned in this book are written as H-1, H-2 and H-3 (you will often see the isotopes written as ^1H, ^2H and ^3H).

Since the hydrogen isotopes are so well known, they have attained their own names. H-2 is called deuterium and H-3 is called tritium. When tritium disintegrates it emits a β-particle with an average energy of only 5.68 keV and maximum energy of 18.6 keV (1 keV equals 1000 electron volts of energy).

In nature, 99.985% of hydrogen is the H-1 isotope. In ordinary water, only one out of 7000 atoms is deuterium. Due to nuclear processes in the atmosphere there are small amounts of tritium. Tritium is widely used in research.

Potassium is another example of an element that has radioactive isotopes. Potassium consists of 93.10% K-39, 6.88% K-41 and 0.0118% of the radioactive isotope K-40. The latter isotope is present because it has a very long half-life of 1.27 billion years. The Earth's crust contains a lot of potassium. In spite of the small fraction of K-40, the radiation from this isotope is quite important. All living organisms contain some radioactive potassium. For example a human being contains, on average, about 60 Bq/kg body weight of K-40 (for more details see Chapter 7).

The Radioactive Series

A radioactive atom is unstable and will eventually eject a particle and/or a photon to attain a more stable state. Certain atoms are still unstable even if radiation has been emitted. Uranium is a typical example shown in Figure 2.3.

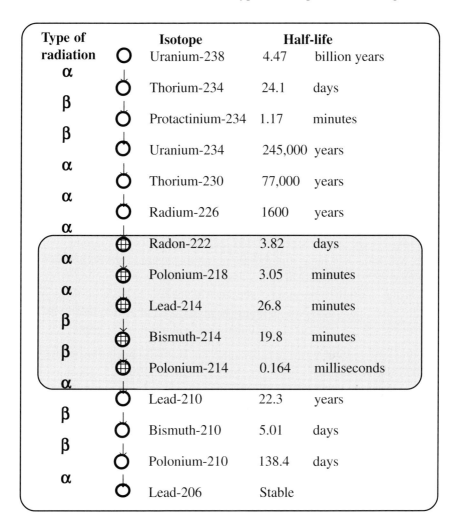

Type of radiation	Isotope	Half-life	
α	Uranium-238	4.47	billion years
β	Thorium-234	24.1	days
β	Protactinium-234	1.17	minutes
α	Uranium-234	245,000	years
α	Thorium-230	77,000	years
α	Radium-226	1600	years
α	Radon-222	3.82	days
α	Polonium-218	3.05	minutes
α	Lead-214	26.8	minutes
β	Bismuth-214	19.8	minutes
β	Polonium-214	0.164	milliseconds
α	Lead-210	22.3	years
β	Bismuth-210	5.01	days
β	Polonium-210	138.4	days
α	Lead-206	Stable	

Figure 2.3. Uranium-radium-series. The start of the series is U-238 and the end point is Pb-206. The first isotope has the longest half-life, 4.47 billion years. Radon and the radon decay products, which you will hear more about, are encircled (notice the short half-lives).

The chemical symbol for uranium is U. The start of this decay series is the isotope U-238 or ^{238}U (see footnote). When this isotope emits an α-particle, it is changed into thorium-234.

$$^{238}\text{U} \Rightarrow {}^{234}\text{Th} + \alpha$$ *This is the way physicists describe the above reaction.*

Th-234 is also unstable and it emits a β-particle, forming a new decay product, protactinium-234. The new product is still not stable and the decay processes continue step by step until Pb-206 is reached. Altogether, 14 disintegrations take place before U-238 ends up as a stable lead isotope (the whole series is shown in Figure 2.3). A series of unstable atoms where one atom changes into another is called a *radioactive family* or simply a *radioactive series*. Altogether, there are 4 naturally occurring radioactive families on Earth. Two of these have almost disappeared and only the uranium–radium series and thorium series are still active.

A radioactive source consists of a large number of unstable atoms. For example, one gram of the iodine isotope I-131 consists of $4.6 \cdot 10^{21}$ atoms. All these atoms will sooner or later emit radiation, but these emissions do not take place simultaneously. It is a statistical process, with one atom decaying every now and then. When one half of the atoms have decayed the source has gone through what is called one "half-life". Not all atoms have decayed after two half-lives, ¼ of the unstable atoms remain (you will learn more about this in Chapter 3).

This uranium–radium series has been present from the beginning of the Earth. The first step in this series has a very long half-life of almost 5 billion years, the present age of the Earth. Thus, we are only now into the second half-life of the uranium–radium series. The two radioactive series that have almost disappeared have done so because the half-lives are much shorter.

The Energy of the Radiation

In order to detect radioactivity and to evaluate the biological effect of the radiation it is important to have information about the energy as well as the type of radiation emitted. The unit used for energy is the *electron volt* (abbreviated eV). By definition, an *electron volt is the energy attained by an electron when it is*

U-238 indicates that the nucleus consists of 238 nucleons. From atomic physics it is known that uranium has 92 electrons, and the nucleus has, consequently, 92 protons. The number of neutrons is therefore 238 − 92 = 146.

accelerated through a voltage gap of 1 volt. The product of voltage and the electron charge (given in Coulombs, C) gives the relation between electron volt and a unit of energy, the joule (J):

$$1 \text{ eV} = 1 \text{ V} \cdot 1.6 \cdot 10^{-19} \text{ C} = 1.6 \cdot 10^{-19} \text{ J}.$$

The electron volt is a very small unit. The energy usually set free by a disintegration varies from a few thousand electron volts (keV) to approximately 6 million electron volts (MeV).

Description of a Radioactive Source

How is a radioactive source described? The intensity of the source depends on the number of atoms that disintegrate per second (i.e. the number of becquerels as defined in Chapter 4). Other parameters are: *type of radiation, half-life,* and *energy of the radiation.* All these parameters can be given by a *decay scheme.* For example, the radioactive isotope Cs-137, which is the most important radioactive waste product from a nuclear reactor, has the decay scheme given in Figure 2.4.

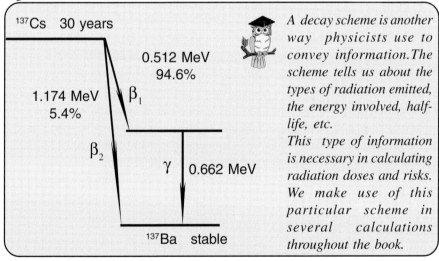

137Cs   30 years

0.512 MeV
94.6%

1.174 MeV
5.4%

β_1

β_2

γ | 0.662 MeV

137Ba   stable

A decay scheme is another way physicists use to convey information. The scheme tells us about the types of radiation emitted, the energy involved, half-life, etc.
This type of information is necessary in calculating radiation doses and risks. We make use of this particular scheme in several calculations throughout the book.

Figure 2.4. A scheme for the disintegration of Cs-137. The state of the nucleus is given by horizontal lines. The atomic number increases left-to-right, Cs is 55 and Ba is 56. The vertical scale is the energy of the nucleus, given in MeV. The vertical distance between the lines indicate the energy difference. This energy is set free by a disintegration, appearing as a β-particle or γ-ray.

The decay scheme shows that Cs-137 is transformed into the stable barium isotope Ba-137. This can take place via two different routes:

1. In 94.6% of the disintegrations a β-particle is emitted with an energy of 0.512 MeV (10^6 eV), followed immediately by a γ-ray with an energy of 0.662 MeV.
2. In 5.4% of the disintegations the stable barium isotope is reached directly by emitting only a β-particle, with an energy of 1.174 MeV.

The decay scheme also shows that the half-life of Cs-137 is 30 years. In addition, one might guess that Cs-137 can be observed by measuring the emitted γ-rays. γ-rays are easy to detect because they are very *penetrating*, a quality that is described at the end of this chapter.

Alpha Radiation

The energy of an α-particle, when it is emitted by a nucleus, is usually a few MeV. Some of the properties which are characteristic of α-particles are:

*   The α-particles from one particular radioactive source have the same energy. For example the α-particles from U-238 always have a starting energy of 4.19 MeV.
*   When an α-particle passes through a material, it rapidly loses energy through numerous collisions with the electrons that make up the atoms and molecules. Because the collisions produce ionizations, a high density of ions are deposited in the material tracing out a linear track. The energy of the α-particle is ultimately dissipated by this large number of low energy interactions and it stops at the end of the track.

The energy deposited per unit length of the track is called the *linear energy transfer* (abbreviated LET). An example is given in Figure 2.5. The range of an α-particle from a radioactive source is very short in animal tissue and in air the range is only a few centimeters. As can be seen from Figure 2.5 the energy loss along the track is not constant but gradually increases toward the end of the track.

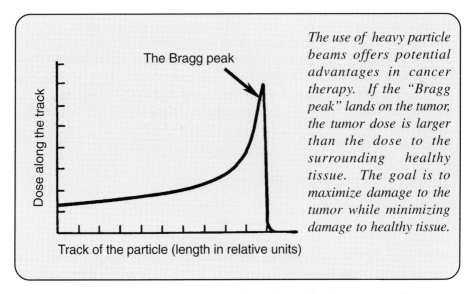

The use of heavy particle beams offers potential advantages in cancer therapy. If the "Bragg peak" lands on the tumor, the tumor dose is larger than the dose to the surrounding healthy tissue. The goal is to maximize damage to the tumor while minimizing damage to healthy tissue.

Figure 2.5. The energy deposition along the track of an α-particle.

Beta Radiation

The energy of a β-particle (a fast electron or positron – the latter is a positively charged electron) is usually much smaller than the energy of α-particles. Furthermore, the energy of the β-particles varies from one disintegration to another. β-particle emission from a source is described by a spectrum of energies. Usually, the maximum energy is given in the decay schemes such as that shown earlier in Figure 2.4.

Consider this in more detail. In a disintegration the nucleus changes from one energy state to another. This change is given as a well defined energy gap. However, the β-particles do not all have the same energy. The explanation is that, together with the β-particle, a tiny neutrally charged particle is emitted. This particle was called a *neutrino* by the Italian physicist Enrico Fermi. (The term neutrino means "the small neutral particle".) The sum of the energies of the electron and the neutrino is equal to the energy gap in the decay scheme.

The average β-particle energy: rule of thumb

The β-particle energy for a source varies from zero up to a maximum. The average energy is approximately 1/3 of the maximum energy.

β-particles are stopped by collisions with electrons in materials in a process similar to the way α-particles are stopped. As a rule of thumb one can say that a β-particle with energy 1 MeV will have a range in water or soft tissue of 0.5 cm. The β-particles from Cs-137 have an average energy of 0.2 MeV. If these particles hit the skin, the penetration into the body would be less than 1 mm. However, if a sufficient number of these β-particles hit the skin, it will be burned.

Gamma Radiation

The energy of a γ-ray is given by the expression:

$$E = h\nu$$

where h is a fundamental constant known as Planck's constant and ν is the frequency of the radiation wave. The radiation can be considered to consist of small energy packages called quanta or photons. The energy of the γ-ray ranges from 0.1 to 1.5 MeV. The cesium isotope Cs-137 emits γ-rays with an energy of 0.662 MeV. The cobalt isotope Co-60 emits two quanta with energies of 1.17 and 1.33 MeV.

Gamma-rays and x-rays are absorbed differently from α-particles. When γ-rays penetrate a material, the intensity of the radiation (I) decreases according to an exponential formula:

$$I(x) = I_o \cdot e^{-\mu x}$$

where x is the depth in the material and μ is the absorption coefficient (μ describes how the radiation decreases per unit length for each type of material). The absorption coefficient has three different components. This is because three processes are involved: *photoelectric effect, Compton scattering* (inelastic scattering) *and pair production.*

Photoelectric effect is a process in which a photon interacts with a bound electron. The photon itself disappears, transferring all its energy to the electron and thereby imparting kinetic energy to the electron. This is the most important absorption process for radiation with an energy less than about 100 keV (which is the type of radiation used in medical diagnostics). The photoelectric effect varies dramatically with the electron density of the absorbing medium. Thus material that contains atoms with high atomic numbers, e.g., the calcium in bone, gives strong absorption due to the photoelectric effect.

Compton scattering is a process in which a photon collides with a bound electron and where the photon energy is considerably greater than the electron binding energy (see Figure 2.6).

After the interaction, the photon continues in a new direction with reduced energy and the electron attains enough energy to leave the atom. We call this electron a *secondary electron*. The Compton process is the most important absorption process for photons with energies from about 100 keV to approximately 10 MeV (the type of radiation mainly used for radiation therapy).

Pair production is a process in which the energy of the photon is used to produce a positron–electron pair. The photon energy must be above 1.02 MeV,

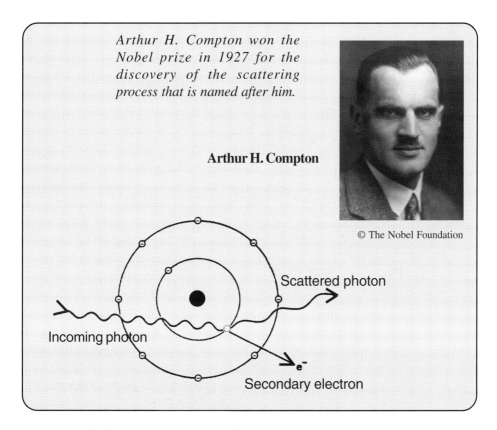

Arthur H. Compton won the Nobel prize in 1927 for the discovery of the scattering process that is named after him.

Arthur H. Compton

© The Nobel Foundation

Scattered photon

Incoming photon

e⁻
Secondary electron

Figure 2.6. The drawing describes the process. The incoming photon interacts with an electron and the result is that the photon is scattered and its energy is reduced. The electron is ejected and becomes a "secondary electron".

the threshold energy for forming two electrons. The process takes place near the atomic nucleus and is the most significant absorption mechanism when the photon energy is above about 10 MeV.

The Penetration of Radiation

When using a gun, the penetration by the bullet depends on the energy of the bullet as well as the composition of the target. For example, a pellet from an air gun will be stopped by a few millimeters of wood but a bullet from a high powered rifle will pass through many millimeters of steel. It is similar with ionizing radiation. There are large differences in penetrating ability depending on the type of radiation (α-, β- or γ-radiation).

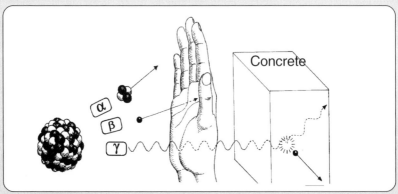

Alpha-particles from radioactive sources have energies up to 6 to 7 MeV with a range in air of only a few cm. In condensed matter the range is much shorter, α-particles will not even penetrate clothing. As long as the α-particle source is outside the body, there is no danger. If, however, the source is inside the body, all the energy is deposited in the body.

It is mainly the heavy elements such as uranium, radon and plutonium which are α-emitters.

Beta-particles with an energy of 1 MeV have a range in soft tissue of approximately 5 mm. The majority of β-particles have an energy far less than 1 MeV. Consequently, almost all β-particles coming from sources in the environment are stopped by clothing.

Gamma radiation has the ability to penetrate tissue and even concrete (figure). For example, 50% of the γ-rays from Cs-137, with energies of 0.662 MeV will penetrate a water layer of about 9 cm. We call this a *half-value layer*. Five half-value layers (less than 0.5 meter of water) will reduce the radiation by 97%. γ-radiation is easy to measure, whether the source is outside or inside the body. Consequently, isotopes emitting γ-radiation are used in medical diagnoses.

X-rays and γ-rays will easily penetrate the human body. This property is utilized when x- and γ-rays are used for diagnostic purposes. α- and β-particles, on the other hand, lose their energy within a short distance and cannot penetrate the body. Because of these penetration properties, γ-radiation is easy to observe whereas α-and β-radiation are more difficult to detect. Thus, special instruments are often needed in order to observe α- and β-rays. The following conclusions can be drawn:

- If a radioactive source is on the ground, such as in a rock, the α- and β-radiation will be stopped by air and clothes. Only γ-rays would penetrate into the body and deliver a radiation dose.

- When a radioactive source is **inside** the body, it is a different situation. α- and β-particles are completely absorbed within a short distance in the tissues, whereas only a certain fraction of the γ-radiation is absorbed. The rest of the γ-radiation escapes and can be observed with counters outside the body. Consequently, if you eat food containing radioactive compounds, they can be easily measured if γ-rays are emitted.

It is possible then to measure the radioactivity that is inside animals and humans who have eaten food containing Cs-137 due for example to fallout from nuclear tests or nuclear accidents. For adults, approximately 50% of the γ-radiation escapes the body and the other half is absorbed by the body. Other important isotopes such as Sr-90 (strontium) and Pu-239 (plutonium) are very difficult to observe since they only emit β-particles and α-particles.

{Advanced Reading}

What is ionizing radiation?

Ionizing radiation is a more precise name for all types of radiation with energy large enough to ionize a molecule (below, M = molecule). Included under this designation are radiation from radioactive sources (α-, β- and γ-rays), x-rays, short wavelength UV, particles from accelerators, particles from outer space, and neutrons.

$$M + radiation \quad\Longrightarrow\quad \begin{array}{ll} M^+ + e^- & \text{Ionization} \\ M^* & \text{Excitation} \end{array}$$

Ionization

How much energy is necessary?

Most atoms and molecules have an ionization energy of 10 eV and more (the unit eV is defined on page 12). Certain molecules in liquids and in the solid state may have an ionization energy as low as 6 eV. This means that UV-radiation with a wavelength below approximately 200 nm (6.2 eV) may cause ionization. Radiation with an energy of 1 MeV has enough energy to yield about 150,000 ionizations if all the energy deposited produces ions.

Secondary electrons

The electron which is ejected from the molecule in an ionization process is called a *secondary electron*. Secondary electrons with a starting energy of 100 eV or more make their own tracks and will ionize and excite other molecules. These electrons are called delta rays.

Excitation

Ionizing radiations not only ionize but can also excite molecules. Excitations are also produced by long wavelength UV and visible light (called *non-ionizing* radiation). An excitation occurs when the molecule attains extra energy. This is done by increasing the vibrational, rotational, or electronic energies of the molecule. These excited states have short life times (less than milliseconds) and sometimes relax back to the ground state by emitting light. Light emission from an excited molecule, called fluorescence and phosphorescence, is a property that is used to measure and characterize ionizing radiation (see Chapter 6).

Excited and **ionized** molecules are very reactive and have short life times. These reactive products represent the starting point for all radiobiological effects, such as cancer. The biological effect increases with the number of ions and excited molecules formed.

Chapter 3

Radioactive Decay Laws

This chapter contains the necessary information for those interested in estimating doses and carrying out risk calculations in connection with radioactive fallout

Half-life

The term half-life was mentioned earlier. So far, this has referred to only the physical *half-life*. When considering health and the environment, *biological half-life* is also used.

> *The half-life is defined as the time elapsed when the intensity of the radiation is reduced to one half of its original value.*

The Physical Half-life

The radiation from a radioactive source will gradually be reduced. The rate of this decay is given by the half-life. It is usually denoted as $t_{1/2}$ but sometimes denoted as t_p for the physical half-life. In an experiment in which the intensity of the radiation is measured versus time, a curve like that shown in Figure 3.1 is observed. The activity of the radiation is given along the vertical axis (100% when the experiment is started) and the time (in half-lives) is given along the horizontal axis.

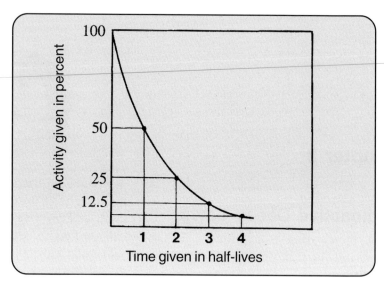

Figure 3.1. The radiation from a radioactive source decreases with time as shown. The curve can be described by an exponential formula. The figure demonstrates the meaning of the half-life.

After one half-life the intensity of the radiation has decreased to 50%. After two half-lives only 25% remains and so on. Each half-life reduces the remaining amount by one half.

The Earth still contains large amounts of naturally occurring radioactive isotopes, such as U-238. For this to occur the half-lives must be very long. We saw in Figure 2.3 that U-238 has a half-life of 4.47 billion years.

The Laws of Radioactive Decay

The activity of a radioactive source (A), i.e. the number of disintegrations per second (becquerel), is given in the following way:

$$A = -dN/dt = \lambda \cdot N \qquad (3.1)$$

λ is the *disintegration constant* and it varies from one isotope to another. N is the number of atoms that, in time, will disintegrate and dN is the change in N during the time interval dt. The negative sign shows that the number remaining is decreasing.

Equation 3.1 shows that when N is larger, the radioactive source is stronger. The difference in activities from one isotope to another is due to the different half-lives which depend on different disintegration constants λ (see equation 3.3).

In order to determine how the the number of atoms (N) decreases with time, the change in N must be summed over time. This is done mathematically by integrating, giving:

$$N = N_o \cdot e^{-\lambda t} \tag{3.2}$$

N_0 is the number of radioactive atoms at time zero (i.e., when the first measurement was made). By substituting a later time (day, year) for t in (3.2) we can solve the equation and determine the radioactivity at the new time.

The two equations (3.1) and (3.2) are very important in order to evaluate risks and radiation doses. These equations are used in the examples presented in Chapter 14.

It was noted above that there is a relation between the half-life ($t_{1/2}$) and the disintegration constant λ. The relationship can be found from equation (3.2) by setting $N = \frac{1}{2} N_0$. This gives:

$$t_{1/2} = \frac{\ln 2}{\lambda} \tag{3.3}$$

where ln 2 (the natural log of 2) equals 0.693.
If the disintegration constant (l) is given, it is easy to arrive at the half-life, and vice versa. In calculations using radioactive compounds one of these two constants must be known.

Biological Half-life

The radioactive isotopes that are ingested or taken in through other pathways will gradually be removed from the body via kidneys, bowels, respiration and perspiration. This means that a radioactive atom *can* be expelled before it has had the chance to decay. The time elapsed before half of the compound has been removed through biological means is called the *biological half-life* and is usually written t_b.

If a radioactive compound with physical half-life t_p $(t_{1/2})$ is cleared from the body with a biological half-life t_b, the *"effective" half-life* (t_e) is given by the expression:

$$\frac{1}{t_e} = \frac{1}{t_p} + \frac{1}{t_b} \qquad \text{or:} \qquad t_e = \frac{t_p \cdot t_b}{t_p + t_b}$$ (3.4)

If t_p is large in comparison to t_b, the effective half-life is approximately the same as t_b.

The biological half-life is rather uncertain compared to the exact value of the physical half-life. It is uncertain because the clearance from the body depends upon sex, age of the individual and the chemical form of the radioactive substance. The biological half-life will vary from one type of animal to another and from one type of plant to another.

Cs-137, having a physical half-life of 30 years, is a good example. It was the most prominent of the radioactive isotopes in the fallout following the Chernobyl accident in the Ukraine. Cesium is cleared rather rapidly from the body and the biological half-life for an adult human is approximately three months and somewhat less for children. Cs-137 has a biological half-life of 2 to 3 weeks for sheep, whereas for reindeer it is about one month.

Due to the fact that the biological half-life for animals like sheep is rather short, it is possible to "feed down" animals, with too high a content of Cs-137, before slaughtering. The animals can simply be fed non-radioactive food for a short period. Another possibility is to give the animals compounds such as "Berlin blue" which is known to speed up the clearance of cesium from the body. The result is a shorter biological half-life.

Some radioactive species like radium and strontium are bone seekers and, consequently, are much more difficult to remove. The biological half-life for radium is long, and if this isotope is ingested, it is retained the rest of one's life.

It is possible to reduce the effects of a radioactive compound by simply pre-venting its uptake. Consider iodine. If people are to be exposed to radioactive iodine, it is possible to add non-radioactive iodine to their food. All iodine isotopes are chemically identical and the body can not discriminate one isotope from the other. There will be a competition between the different isotopes. If the amount of non-radioactive iodine is larger than the radioactive isotope the uptake of radioactivity is hindered. This kind of strategy can also be used to decrease the biological half-life.

C-14 used as a biological clock

Radioactive carbon (C-14) has a half-life of 5730 years. In spite of this rather "short" half-life, C-14 is a naturally occurring isotope. It is created continuously in the atmosphere when neutrons (originating from cosmic radiation) interact with nitrogen atoms. Carbon exists in the atmosphere as a component of carbon dioxide and enters the biosphere when plants utilize carbon dioxide in photosynthesis. All biological systems, plants, animals and humans contain a certain level of C-14. The uptake of C-14 stops with death. From then on, the radioactivity will decrease according to the curve below. The percent decrease can be used to determine the age of organic materials, such as wood. The use of C-14 to determine age is called carbon dating.

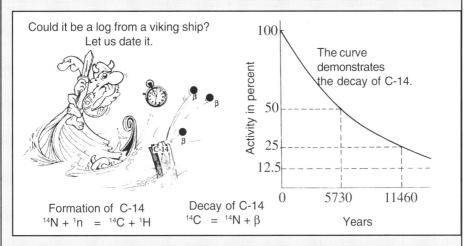

Could it be a log from a viking ship? Let us date it.

The curve demonstrates the decay of C-14.

Formation of C-14
$$^{14}N + ^1n = ^{14}C + ^1H$$

Decay of C-14
$$^{14}C = ^{14}N + \beta$$

C-14 emits β-particles. Since the energy is small (max. 156 keV) and since the number of disintegrations is small, the usual C-14 dating method has several problems. It requires rather large samples (many grams) in order to yield enough radiation to provide a high degree of certainty in the age determination.

Carbon dating is based on the measurement of β-particles from C-14 atoms that disintegrate during measurement. For each becquerel there are 260 billion C-14 atoms (see example 1 in Chapter 14). Consequently, if one could observe the **total amount of C-14 atoms** in a sample (not only those disintegrating per second), both the sensitivity and the age determination would be increased.

The American physicist Louis Alvarez developed a dating method based on this principle. He used a very sensitive instrument, called a mass spectrometer, that detects C-14 atoms based on their atomic mass. He measured the total number of C-14 atoms, not only those that disintegrated during the observation period. With this technique, it was possible to date very small samples (a few milligrams).

Nuclear tests that in the past were conducted in the atmosphere released neutrons that increased the formation of C-14. Because of the 5730 year half-life, we will have these extra C-14 atoms for a long time.

Radio-ecological Half-life

Radio-ecological half-life is less precise than the physical and biological half-life. Consider a region which has been polluted by a radioactive isotope (for example Cs-137). Part of the activity will gradually sink into the ground and some will leak into the water table. Each year, a fraction of the activity will be taken up by the plants and subsequently ingested by some of the animals in the area.

Radio-ecological half-life is defined as the radioactive half-life for the animals and plants living in the area. It varies for the different types of animals and plants. Knowledge in this area is limited at present, but research carried out after the Chernobyl accident has yielded some information.

Here is one example from that accident. It describes the radioactivity in trout in a small lake in the middle of Norway. The measurements were begun the spring of 1986 and carried out for a 4 year period. The results are given in Figure 3.2.

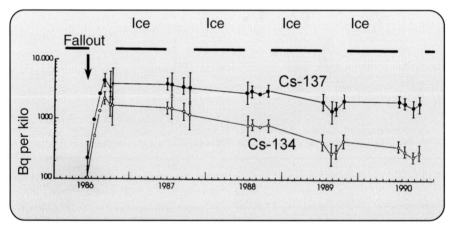

Figure 3.2. Radioactive trout after the Chernobyl accident given in becquerel per kg (note the logarithmic scale). The lake is covered with ice each year, as given by the heavy lines. (Courtesy of Anders Storruste, Inst. of Physics, Univ. of Oslo)

Remember that to determine a half-life we need to use an exponential equation (see equations 3.2 and 3.3). But the data shown in the above figure does not fit an exponential function. Therefore, it is impossible to arrive at a single ecological half-life. However, as Figure 3.2 indicates, the half-lives are approximately 3.0 years for Cs-137 and 1.3 years for Cs-134. It is important to note that these ecological half-lifes are significantly shorter than the respective physical half-lifes, 30 years for Cs-137 and 2 years for Cs-134.

Chapter 4

Activity and Dose

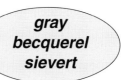

gray
becquerel
sievert

Activity in Becquerel

When an atom disintegrates, radiation is emitted. If the rate of disintegrations is large, a radioactive source is considered to have a high activity.

The unit for the activity of a radioactive source was named after Becquerel (abbreviated Bq) and is defined as:

> 1 Bq = 1 disintegration per sec.

In a number of countries, the old unit, the curie (abbreviated Ci and named after Marie Curie) is still used. The curie-unit was defined as the *activity in one gram of radium*. The number of disintegrations per second in one gram of radium is 37 billion. The relation between the curie and the becquerel is given by:

> 1 Ci = $3.7 \cdot 10^{10}$ Bq

The accepted practice is to give the activity of a radioactive source in becquerel. This is because Bq is the unit chosen for the system of international units (SI units). But one problem is that the numbers in becquerel are always very large. The opposite holds true when a source is given in curies. For example, when talking about radioactivity in food products, 3,700 Bq per kilogram of meat is a large number. The same activity given in Ci is a really small number, 0.0000001 curie per kilogram.

Most people are used to measuring the amount of a substance in kilograms or liters, not in becquerels or curies. Using the best balances in the world one can accurately measure down to about one microgram of substance (10^{-6} gram). By using our knowledge of radioactive detection, amounts more than a million times smaller than this can be measured. Radioactive sources as small as 10 Bq to 100 Bq can be readily measured; this corresponds to only about 10^{-14} gram.

Some examples will help develop some familiarity with the Ci and Bq units. Around 1930, radium sources were introduced in many hospitals and used for cancer therapy. Radium was rather expensive and the sources used for teletherapy (see page 29) were 2 to 10 Ci. With these samples placed only about 10 cm from the skin, a treatment lasted for about half an hour. Small samples of radium (in the milligram range) were used for brachytherapy (see below). In some cases, needles with radium were melted into solid paraffin and placed directly on the skin of a patient.

Today, sources used for radiation therapy usually consist of Co-60 or Cs-137 and the total activity is about 5,000 Ci ($185 \cdot 10^{12}$ Bq or 185 TBq. 1 TBq is 10^{12} Bq). Much stronger sources (on the order 100,000 Ci) are used for sterilizing medical equipment. In Chapter 10, an example is given where a man accidentally received a lethal dose from a Co-60 source of 65,000 Ci (2,400 TBq).

It should be noted that the content of radioactivity in certain ordinary compounds can be surprisingly large. For example, if a person goes to a store and buys 1 kg of potassium hydroxide (KOH), they have purchased a radioactive source containing K-40 with an activity of 21,800 Bq.

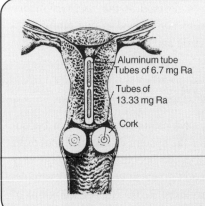

Aluminum tube
Tubes of 6.7 mg Ra

Tubes of
13.33 mg Ra

Cork

An example of brachytherapy

This drawing from 1931 demonstrates the treatment of a cervical cancer by using small needles containing radium.

Cork containers are used for the vaginal sources and a cylindrical tube for the uterine sources. As can be seen from this drawing, the total amount of radium was about 40 milligram. This means that the activity of the radioactive source was 40 mCi or approximately 1.5 billion Bq.

(From R.F. Mould (1993), Fig. 22.18, p. 177; with permission from R.F. Mould)

Radium therapy

In the period 1930 to 1960, radium was widely used for cancer treatment. Several hospitals around the world, which used radium extensively, were called Radium Hospitals. Since radium was rather expensive in those days (approximately $30,000 per gram) the total amount in each hospital was limited. Two different types of use were introduced; teletherapy (with sources of 2 to 10 gram), and brachytherapy (surface, interstitial and intercavitary) with milligram sources.

Teletherapy

To the right is shown an old so-called "radium cannon" used for therapy. The source consisted of 3 grams of radium. For treatment, the source was removed from the lead container and moved down into the tubes. The source-to-skin distance was approximately 10 cm, and a treatment lasted about 45 minutes for a dose of 2 Gy. This was a long time for a patient given that they could not move out of position during treatment.

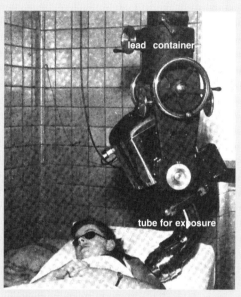

Courtesy of Norwegian Radium Hospital

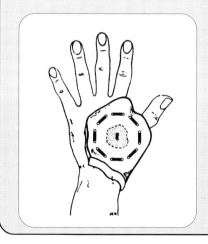

Surface therapy

To the left is shown an example of how small radium needles were arranged in a surface treatment. Very often solid paraffin was formed over the part of the body to be treated. Then needles containing a few milligram of radium were "melted" into the paraffin-mold. A surface treatment with radium usually lasted a couple of days.

(From R.F. Mould (1993), Fig. 20.10, p. 131; with permission from R.F. Mould)

Specific Activity

Specific activity is the activity per mass or volume unit. For example, the radioactivity in meat is given as Bq/kg. For liquids the specific activity is given in Bq/l and for air and gases the activity is given as Bq/m^3.

In the case of fallout from a nuclear test or accident, the activity on surfaces can be given either as Bq/m^2 or as Ci/km^2. Both are used to describe radioactive pollution. The conversion between them is:

$$1 \; Ci/km^2 = 37,000 \; Bq/m^2$$

A great deal of information must be considered to calculate radiation doses and risk factors associated with these specific activities. The information must include the specific activity along with the various types of isotopes, their energies, physical and biological half-lives and methods of entry into the body. After considering all of these factors and calculating the dose, a determination of medical risk can be calculated.

Radiation Dose

A strong radioactive source represents no risk as long as it is isolated from populated environments. It is only when people are exposed to radiation that a radiation dose is delivered.

It is very important to distinguish between the activity of a radioactive source (measured in becquerels) and the radiation dose which may result from the source. The radiation dose depends on the location of the source with regard to those exposed. Furthermore, the radiation dose depends upon the type of radiation, such as whether it is α-, β- or γ-rays and the energy of the radiation.

Although people can neither see nor feel radiation, it is known that radiation deposits energy in the molecules of the body. The energy is transferred in small quantities for each interaction between the radiation and a molecule and there are usually many such interactions. For anything that is irradiated, the temperature rises. Additional radiation increases the temperature further. The temperature increase occurs because the radiation energy is transformed into heat. Even though it is generally very difficult to detect the rise in temperature, the realization that heat is generated by radiation is a key element in understanding the concept of *radiation dose.*

Radiation dose measures the amount of energy deposited in an irradiated compound

1 Gy = 1 joule absorbed energy per kg

Dose Units and Their History

In the course of the 100 years of dealing with ionizing radiation, several different dose units have been used. Some of these units are still used in different countries. It is useful, therefore, to consider some of these units and to see the relations between the old units and the gray unit (Gy).

- **Skin erythema dose**

It was discovered early that radiation exposure resulted in reddening of the skin. For a long period this reddening was used to quantify the radiation. This was called the *skin erythema dose*. This unit was quite uncertain since the reddening of the skin varied from one person to another. Another drawback was that the reddening appeared some time *after* the exposure.

In the case of ultraviolet radiation, this dose unit (along with the attending uncertainties) is still in use. The smallest UV-dose resulting in the reddening of the skin is called **MED**, which is an abbreviation of *minimum erythema dose*.

- **The roentgen unit**

People who worked with radiation circa 1920 began searching for a more precise dose unit and in 1928, the *roentgen unit* (abbreviated R) was adopted. This unit can not be used for the dose itself since it is actually a measure of radiation exposure, i.e. the ionization of air molecules.

An exposure of 1 R means the amount of x- or γ-radiation that results in

$2.58 \cdot 10^{-4}$ coulomb per kg of ions generated in air.

To calculate the radiation dose (in Gy) from an exposure of 1 R depends on the energy of the x- or γ-radiation and the composition of the irradiated material. For example, if soft tissue is exposed to γ-radiation of 1 R, the radiation dose will be approximately 9.3 milligray (mGy).

- **The rad unit**

In 1953, the dose unit *rad* was developed. This is an abbreviation for *radiation absorbed dose* and is defined as:

The amount of radiation which yields an energy absorption of 100 erg per gram (i.e. 10^{-2} joule per kg).

The rad unit is still used in several countries.

$$1 \text{ gray} = 100 \text{ rad}$$

In this book, the gray is used most of the time. But use of the rad is difficult to avoid due to its pervasive use in the older literature. The SI-system of units uses the gray.

Equivalent Dose

When a biological system is exposed to ionizing radiation, molecules are ionized and excited. For a particular material, all types of radiation yield the same kinds of damaged molecules. However, there are differences with regard to the distribution of ionized and excited molecules along the track of a particle or photon.

Consider a very powerful microscope that makes it possible to "see" the molecules in a system. If a microscope is used when the system is exposed to radiation, differences will be observed in the distribution of ionizations produced by the radiation (see Figure 4.1). X-rays, β-particles and γ-rays strike the molecules rather sporadically – as indicated in the upper part of Figure 4.1. The small dark dots represent ionized molecules. For α-particles the situation is different. The ionizations are deposited along a single linear track. The molecules in the center of the track (the track core) are ionized, or hit. Sparse ionization along the γ-ray track is due to a low *linear energy transfer* (LET) while the dense ionization of the α-particle is due to high LET.

Observation: Even though the number of ionized molecules, and consequently the energy deposition, within the two circles is the same (the absorbed dose is the same), the distribution of the energy deposition (i.e., the LET) is different.

The biological effect depends upon the LET of the radiation, i.e., the distribution of the absorbed energy.

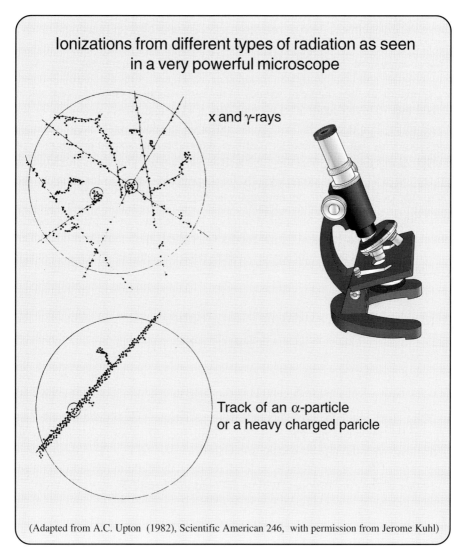

Ionizations from different types of radiation as seen in a very powerful microscope

x and γ-rays

Track of an α-particle
or a heavy charged paricle

(Adapted from A.C. Upton (1982), Scientific American 246, with permission from Jerome Kuhl)

Figure 4.1. The Figure indicates how the distribution of absorbed energy in a system (for example an animal cell) might look after different types of radiation have passed through. The upper circle (field of view) contains tracks produced by x- and γ-ray absorption and the lower circle contains the track of an α-particle. Each dot represents an ionized molecule. The number of dots within the two circles is the same, indicating the same radiation dose. However, note that the distribution of dots (ionizations) is quite different. The top is an example of low LET (linear energy transfer) and the bottom is an example of high LET.

For biological effects, such as cell death, cancer induction, and genetic damage, the effect is larger when the radiation energy is deposited within a small region. This is calculated by introducing a *radiation weighting factor* (w_R).

w_R is related to the relative efficiency of the radiation in producing a biological effect, and is given relative to high energy x-rays and γ-rays where w_R is set equal to 1. For α-particles, neutrons and other particles, w_R is larger than 1. This indicates that most biological effects depend on the spatial distribution of energy along the track (i.e. the LET or Linear Energy Transfer). In radiation protection, a w_R of 20 is currently used for α-particles where as for neutrons the factor varies from 5 to 20, depending upon their energy.

When the physical dose, measured in gray, is multiplied by w_R, the biological effective dose is calculated. This product is called the equivalent dose. The unit for equivalent dose is the sievert (abbreviated Sv), named after the Swedish scientist R.M. Sievert.

$$H = w_R \cdot D$$

H is the equivalent dose in Sv and D is the dose in Gy.

In order to evaluate the biological effect of radiation we apply the sievert. The crux of the matter is the value of w_R. A number of experiments have been performed with the aim of getting more information on w_R. This parameter is frequently of crucial importance. A good example, the radiation dose from radon and other α-emitting isotopes, will be covered in Chapter 7.

Effective Equivalent Dose

In some cases, only a part of the body is irradiated. For example, mainly the lungs are involved in the case of radon and radon decay products. Different organs and types of tissue have different sensitivities with regard to what is termed the *late effects* of radiation. Late effects are biological responses that are only observed after a substantial amount of time has passed, often years. Induction of cancer is a late effect. In order to compare the risk for late effects of different types of radiation, the so-called *effective dose* is used.

Rolf M. Sievert

L. Harold Gray

Reproduced with permission from the Radiation Research Society

Reproduced with permission from the Radiation Research Society

$$\text{sievert} = \text{gray} \cdot w_R$$

Rolf M. Sievert (1896–1966) was educated at the Technical High school in Stockholm and at the University of Uppsala in Sweden. He was interested in radiation physics and he constructed an ionization chamber for depth dose measurements.

He played a significant role in establishing the international committees ICRP (International Commission on Radiological Protection) and UNSCEAR (United Nations Committee on the Effects of Atomic Radiation). He served for several years as director of the Swedish National Institute of Radiation Protection.

Sievert is honored for his work in dosimetry (the measurement of absorbed dose) by naming the unit for equivalent dose *sievert* (abbreviated Sv).

L. Harold Gray (1905–1965) was one of the great pioneers in radiation biology. He was educated as a nuclear physicist and worked in the Cavendish laboratory at Cambridge.

He became interested in the effect of oxygen on radiosensitivity. Cells with a low content of oxygen (hypoxic cells) are sensitive to radiation as normal cells. This behavior has caused problems for the treatment of cancer because most tumors contain regions with hypoxic cells. Gray is well known for fundamental contributions in this area.

When the new SI-system was introduced the radiation dose unit was called *gray* (abbreviated Gy).

If one part of the body (e.g., the lungs) receives a radiation dose, it represents a risk for a particularly damaging effect (e.g., lung cancer). If the same dose is given to another organ it represents a different risk factor.

> *It is possible to calculate a dose given to the whole body that yields the same risk as that from the much larger dose given to one particular organ.*

This calculated dose is called *the effective dose* (often shortened to simply *the dose*) and is designated *E*. It is defined in the following way:

$$E = w_1 H_1 + w_2 H_2 + \ldots\ldots$$

where w_1 represents a weighting factor for organ 1 and H_1 is the equivalent dose (given in Sv) for organ number 1. The weighting factor represents the sensitivity of a particular organ and Table 4.1 gives the different weighting factors suggested for radiation protection work by the ICRP. Equivalent dose and effective equivalent dose have a meaning only when considering late effects such as cancer and leukemia.

Other Dose Units

In addition to the units already defined, there are a number of other concepts used in radiation protection and these, unfortunately, are frequently a source of confusion. Radiation biologists are interested in the mechanisms that result in the observable macroscopic effects on biological systems, whereas radiation protection authorities are interested in guidelines for large population groups. Here is a brief overview of the concepts used when working in radiation protection.

• **Collective dose**
The collective dose is the sum of all individual doses in a group of people. It can be obtained by the product of the average individual dose with the number of people in the group. For the collective dose the unit used is person-sievert (person-Sv). For reasons to be discussed later in this book, the use of collective dose is questionable in risk analyses. Nevertheless, it is used by authorities for risk analysis in many countries (see Chapter 11).

Table 4.1. Tissue weighting factors used in radiation protection work

Organ	Weighting factor
Gonads	0.20
Bone marrow (red)	0.12
Lungs	0.12
Stomach	0.12
Colon	0.12
Bladder	0.05
Breast	0.05
Liver	0.05
Thyroid gland	0.05
Esophagus	0.05
Bone surfaces	0.01
Skin	0.01
The rest	0.05
TOTAL	1.00

- **Committed equivalent dose**

When a radioactive compound enters the body, the activity will decrease with time, due both to physical decay and to biological clearance, as noted earlier. The decrease varies from one radioactive compound to another. Accumulated dose over a certain period of time, usually 50 years, is called the committed equivalent dose.

How little radioactivity can be observed?

The instruments developed to measure radioactivity are very sensitive, and extremely small amounts can be observed. Can you believe that it is possible to observe 1 gram of a compound after it has been distributed all over the world? Given even the best balances you would very soon realize that it is impossible, but if the compound is radioactive you may have a chance. For example, if 1 gram of I-131 is distributed all over the world (an area of $5.1 \cdot 10^{14} \, m^2$), the specific activity would be approximately 10 Bq per m^2. You can check this calculation using the fact that 1 gram of I-131 has an activity of $4.59 \cdot 10^{15}$ Bq (see exercise 2 of Chapter 14).

Listen all of you!
Do you think you can measure
1 gram of I-131 if it is distributed
all over the world?

It is possible to observe a radioactive source of 10 Bq per m^2. This is indicative of the sensitivity of equipment used for measuring radioactivity. Other radioactive isotopes may require larger amounts. The reason is that the activity is closely connected to the half-life. For I-131, with a half-life of 8 days, 1 gram can be observed, whereas for isotopes with longer half-lives larger amounts are required.

For example Cs-137, with a half-life of 30 years and with 1 gram distributed all over the world, would have a specific activity of only 0.007 Bq per m^2. This would not be detectable.

The total amount of Cs-137 released in the Chernobyl accident was about 12 kg. If this amount was distributed evenly around the world the activity would be 83 Bq per m^2. In some regions in Scandinavia the fallout reached 100,000 Bq per m^2. Near the reactor the amount was more than 10 times larger.

Chapter 5

Artificial Radioactive Isotopes

The Discovery

In 1934, Irene Joliot Curie (Marie Curie's daughter) and her husband, Frederic Joliot, succeeded in making a radioactive isotope that does not occur in nature. They bombarded an aluminum plate with α-particles from a natural radioactive source, and when they removed the α-particle source, it appeared that the aluminum plate emitted radiation with a half-life of approximately 3 minutes. The explanation was that the bombardment had resulted in a nuclear reaction. The α-particle penetrated the aluminum nucleus and changed it into phosphorus by emitting a neutron. The new phosphorus isotope was radioactive and was responsible for the observed radiation. Its designation is P-30.

Irene and Frederic Joliot Curie

Berkeley National Laboratory, University of California Berkeley, courtesy AIP Emilio Segrè Visual Archives

This nuclear reaction is written as follows:

$$^{27}Al + \alpha \Rightarrow {}^{30}P + n.$$

The neutron emitted can be observed as long as the bombardment takes place, but disappears immediately when the α-source is removed. However, the phosphorus isotope is radioactive and emits a positron with an energy of 3.24 MeV (which can easily be measured) and a half-life of 2.50 minutes.

In the mid 1930's, several laboratories had developed equipment to accelerate protons and α-particles to high energies. When these elementary particles were used as projectiles to bombard stable atoms, new isotopes were formed. Some of these isotopes were radioactive.

Another very efficient particle used in these experiments was the neutron. This particle has no charge and will consequently not be influenced by the electric field around the atomic nucleus. The neutron readily penetrates the atom, forming new isotopes. Reactors are excellent sources of neutrons and are used for the production of radioactive isotopes needed for biomedical research and the treatment of disease.

The number of artificial isotopes increased rapidly in the years after 1934. By 1937, approximately 200 isotopes were known, in 1949 the number was 650 and today more than 1,300 radioactive isotopes have been produced.

Fission

After Chadwicks discovery of the neutron in 1932, a large research effort was started in order to make and identify the isotopes formed when neutrons penetrate various atomic nuclei. In 1938, it was observed that one of the largest atoms, uranium, disintegrates in a dramatic way. This unstable nucleus splits into two large fragments. This reaction is called *fission* (see opposite page).

The splitting of a heavy atomic nucleus, such as U-235, occurs because of intrinsic instabilities. The nucleus can exist in a variety of energy states and there are numerous pathways by which the nucleus emits energy and creates new products. More than 200 fission products from uranium are known. The products formed can be divided into two groups, one "heavy" group with an atomic weight of about 140 units and one "light" group with an atomic weight of 90. This is illustrated for U-235 in Figure 5.1.

A large amount of energy is released in fission. Most of the energy is released directly during the process of fission but a small amount is released at a later stage by those fission products that are radioactive. Most fission products have short half-lives. From an environmental point of view, Cs-137 and Sr-90 are the most important fission products of U-235. They each have a half-life of about 30 years, which is important with regard to storage and disposal of these products. The fission process leads to three different types of radioactive isotopes: *fission products, transuranic elements, and activation products.*

The history of fission

Much of the work on understanding fission occurred in the 1930s. Several laboratories in Europe were engaged in research where heavy atoms such as uranium were bombarded with neutrons.

In 1934 **Enrico Fermi** bombarded uranium and observed β-particles. He interpreted this as an absorption of a neutron and that the altered nucleus emitted a β-particle forming a transuranic element.

Shortly after Fermi's work was published in *Nature*, the German chemist **Ida Noddack** published a paper called *"On element 93"*. She was critical of Fermi's work and suggested that when a heavy atom was hit by a neutron it was a possibility that it can fragment into larger units. Although this interpretation was correct, few listened to her. At The Kaiser Wilhelm Institute in Berlin **Otto Hahn, Lise Meitner** and **Fritz Strassmann** worked with uranium and neutron bombardment. They found, upon chemical analysis, that a compound similar to barium was formed (see inset above right). In Paris, **Irene Joliot Curie** observed a compound similar to lanthanum. We know today that when barium emits a β-particle, lanthanum is formed.

Lise Meitner, a Jewish scientist, had to flee from Germany in the summer of 1938. She came to Sweden with the help of Niels Bohr and was supported by the Nobel Society.

Hahn and Strassmann continued the experiments in Berlin and showed in an experiment on the 17th of December 1938, that it **was not possible to separate barium from the compound formed** when uranium was hit by neutrons. On the 19th of December, Otto Hahn wrote a letter to Meitner about the latest results. He ended his letter with the sentence: *"Perhaps you can suggest some fantastic explanations"*.

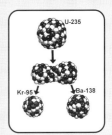

Lise Meitner travelled to Kungälv (just north of Göteborg) to spend the December holidays together with some of her family. Here she met her young nephew **Otto Frisch** who worked with Bohr. One day they sat on the trunk of a tree discussing this phenomenon. Using the nuclear model of Bohr (the so-called liquid drop model) as a basis, they calculated that if a neutron penetrated the nucleus, it could set up oscillations that would split the atom – fission was possible!

Otto Hahn

May be it can be said that the atomic age started on a timber log north of Göteborg during the December holidays of 1938. Hahn and Strassmann published their experiments in the German Journal *Naturwissenschaften*. Meitner and Frisch published their theoretical calculations in the British Journal *Nature* and Bohr let the news explode at a conference in the United States in January 1939.

Lise Meitner

Otto Hahn, A Scientific Autobiography,
Charles Scribner's Sons, New York, 1966,
courtesy AIP Emilio Segrè Visual Archives

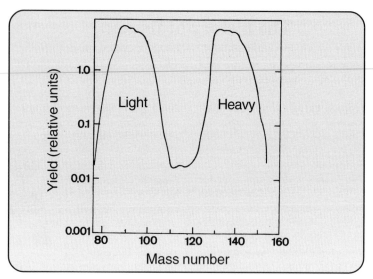

Figure 5.1. Nuclear fission of U-235 results in two new elements. A range of possibilities for fission exists and in each case the two elements are not equal in mass. The average mass of the lighter element is 90 atomic units and of the heavier element 140 atomic units. The distribution is shown by the graph. (Note that the vertical axis is logarithmic.)

Table 5.1. Some important fission products

The Light Group			**The Heavy Group**		
Isotope	Symbol	$t_{1/2}$	Isotope	Symbol	$t_{1/2}$
Krypton-85	Kr-85	10.7 yr	Tellurium-129m	Te-129m	33.6 d
Strontium-89	Sr-89	50.5 d	Iodine-131	I-131	8.04 d
Strontium-90	Sr-90	29.1 yr	Xenon-133	Xe-133	5.3 d
Yttrium-91	Y-91	58.5 d	Cesium-137	Cs-137	30.0 yr
Zirconium-95	Zr-95	64 d	Barium-140	Ba-140	12.7 d
Technecium-99	Tc-99	213,000 yr	Praseodymium-143	Pr-143	13.6 d
Ruthenium-103	Ru-103	39.3 d	Neodymium-147	Nd-147	11.0 d
Ruthenium-106	Ru-106	368 d	Promethium-147	Pm-147	2.6 yr

1. Fission products

There are a large number of fission products, and some of the most important ones are given in Table 5.1. In the first period after fission occurs isotopes with short half-lives dominate, i.e. Zr-95 and I-131. Later, Sr-90 and Cs-137 are predominant.

2. Transuranics

Transuranics are elements with an atomic number larger than 92 (uranium). Most transuranics are made in accelerators when heavy atoms such as uranium are bombarded with neutrons or small charged atoms. Plutonium-239 is a transuranic which is formed when U-238 absorbs a neutron and subsequently emits two β-particles. In a reactor this process can produce large amounts of plutonium.

In addition, the transuranics usually are α-particle emitters, whereas the fission products are β-particle emitters. This implies that when the trans-uranics come into the body, through inhalation or ingestion, they deposit all their energy within the body. And as we learned, the α-particle has a radiation weighting factor (w_R) of 20. Thus, transuranics are a health concern.

3. Activation products

The third type of radioactive isotopes produced in combination with reactors and nuclear weapons are activation products. These radioactive isotopes are formed when stable isotopes are bombarded by neutrons. Co-60 is an activation product formed when Co-59 absorbs a neutron. Likewise, Cs-134 is formed from Cs-133 by neutron capture. The materials around a reactor or a nuclear bomb explosion can be made radioactive by neutron capture.

If a nuclear bomb detonates in the atmosphere, large numbers of neutrons will be released. They can then react with stable atoms such as N-14. In this way the radioactive isotope C-14 is formed. If the detonation takes place just above the ground, large amounts of materials (earth, rocks, building materials, etc.) will be activated and vaporized in the "fireball" which is formed. Such nuclear tests yield large amounts of radioactive fallout.

Plutonium

Few radioactive isotopes have attracted more interest than plutonium. Several plutonium isotopes exist. There are Pu-238 (half-life 87.7 years), Pu-239 (24,400 years), Pu-240 (6,570 years) and Pu-241 (14 years). Pu-239 is the best known and its environmental impact is heavily debated. It is formed in a reactor (see the illustration below) after neutron bombardment of U-238. Pu-239 sits in the spot light because it is not only a cost effective fuel used in fission reactors but also it is a key ingredient in nuclear weapons.

Pu-239 production starts with a neutron absorption by U-238. U-239 and subsequently Np-239 are unstable, and both emit β-particles. The final product is Pu-239.

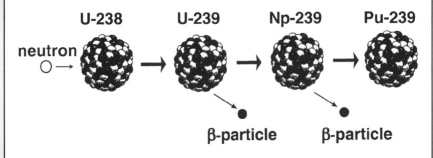

With regard to the environment, Pu-239 is the most important of the plutonium isotopes. It emits an α-particle with an energy of 5.15 MeV. In air, these α-particles have a range of a few cm, whereas in tissue the range is less than one mm. The following important conclusion can be made from this: *plutonium has a minor influence on a person's health when it is outside the body since the emitted α-particles will not enter the body.* On the other hand, if plutonium enters the body all the emitted α-particles will deposit their energy within the body.

Large amounts of plutonium have been released to the atmosphere due to atmospheric nuclear tests. It is estimated, that for the period 1945 to 1974, approximately 400,000 Ci or $1.5 \cdot 10^{16}$ Bq of plutonium were released corresponding to 6.5 tons. As a consequence, plutonium is now found in nature. The fallout of plutonium is approximately as fast as that for strontium (Sr-90). The ratio between plutonium and strontium fallout has been rather constant since the large weapon tests ended in 1963. Calculations show that the total fallout on the Northern hemisphere is approximately 50 Bq per square meter.

Plutonium, which enters the body via the food chain, represents a small radiation problem since only 30 ppm is absorbed in the blood from the intestine. However, the plutonium which enters the body via inhalation (such as those attached to dust particles in the air) presents a more serious problem. See also Chapter 13.

Activation Analysis

The activation of certain materials by neutron irradiation is used as an elegant analytical method for identifying chemical species. When a compound is irradiated with neutrons, many elements are activated and become radioactive. The radioactivity can be measured easily and the properties of the radiation can be used to identify an element. Thus, it is possible to observe the presence of tiny amounts of an element that would be undetectable by other analytical methods.

An archaeologist can also obtain important information from activation analysis in order to determine the properties of old coins, pieces of ceramic pots, and other relics. The method has the advantage that it does not destroy the sample.

Criminologists use activation analyses in the solution of criminal cases. For example, activation analysis showed that the hair of Napoleon contained arsenic. This raises the possibility that he was murdered. Indeed the arsenic could have been introduced intentionally but it also may have come from his environment. At that time, arsenic was used in wall coverings and could have been picked up by touch or given off into the atmospherre.

It is also interesting that determining the composition of the moon was assisted by activation analyses. Rocks, brought back to Earth by the astronauts, were bombarded by neutrons, forming radioactive products. The subsequent radioactive emisions were then used to identify elements in the moon rocks.

Did you know that the composition of the moon was determined, in part, by activation analysis?

Parameters to be determined

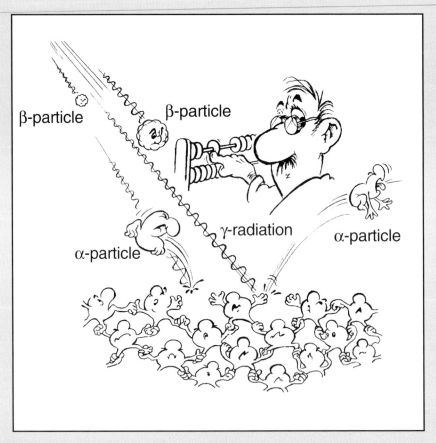

This illustration demonstrates some of the problems connected with the measurement of the radiation from radioactive sources.

A counting system is needed, albeit a little more advanced than the abacus shown here. The most important parts of the equipment are the "eyes which see" the radiation. These "eyes" detect ionizations or scintillations and must be able to separate the different types of radiation (α, β, or γ). In addition, information about the energy of the radiation is needed.

How this is done, i.e., what we use for eyes, is discussed in the next chapter.

Chapter 6

The Measurement of Radiation

Introduction

In this chapter, the equipment and methods used for measuring ionizing radiation will be discussed. There are two types of instruments: counting equipment (used to determine the number of becquerels and the radiation quality) and dosimeters (used to determine the radiation dose). Both types of equipment require that *the radiations result in observable changes in a compound* (whether gas, liquid or solid).

Measuring equipment consists of two parts that usually are connected. The first part consists of a *sensitive volume,* consisting of a compound that experiences changes when exposed to radiation. The other component is *a device that converts these changes into measurable signals.*

The qualities of radiation that we want to measure are:

- **Activity.** The activity or strength of a source is measured in becquerel. The concentration is usually given in Bq/kg (solid) or Bq/m^3 (liquid or gas). In considering pollution of an area Bq/m^2 is used. Some countries still use Ci/km^2.

- **Type of radiation.** It is important to distinguish between α- or β-particles, x- or γ-rays and neutrons.

- **Energy.** The energy is usually measured in the unit electron volt (eV). The energy of the particles or photons is important.

- **Dose** is the absorbed radiation energy. Note that the absorbed energy is different from the energy of the emitted particles (see above). The preferred unit is the gray.

Measuring equipment can vary in size and price. The equipment chosen depends on the purpose of the measurement, the sensitivity desired, and the precision necessary. The simplest type of measurement is to observe the amount of radiation hitting the sensitive volume or detector of the counter. If the sensitive volume is covered with plates of different thicknesses and composition, information may be obtained on the type of radiation. These instruments may be used to monitor radiation areas.

Exact measurements of radiation energy are more complex and different equipment is required. The energy of the radiation may be used to identify a particular isotope. If several isotopes are mixed together (for example, fission products), it is possible to identify the separate isotopes by an accurate determination of the energy of the particles or photons emitted (see Figure 6.4). For that purpose, equipment is needed which has *good energy resolution*.

X-and γ-rays that hit the sensitive volume may pass through it without being absorbed. Consequently, the sensitive volume must be large enough to absorb sufficient amounts of this type of radiation. The greater the energy, the larger the volume. If a significant fraction passes through the sensitive volume without interacting, and therefore without detection, the equipment has a *low efficiency*.

If the sensitive volume contains heavy atoms, the possibility of interaction, and thus the probability of being detected, is greater. Heavy atoms are most effective in stopping x-, γ-, α-, β-, and cosmic rays. When heavy charged particles hit a metal box, they create an electric current that can be measured.

On the next page are given 7 different physical effects that are utilized to measure radiation. Ionization, scintillation, semiconductors and film are extensively used to observe individual particles. All of them can be used to determine radiation doses. In addition to these effects we may also use biological changes, such as chromosome aberrations, for dose determination. For the sake of simplicity, it is important to have a response that is directly proportional to the dose, i.e., the dose–effect relationship should be linear.

Detection Instruments

In this section we describe a number of different types of radiation measuring instruments. Consider the different physical events that can be utilized to make a measurement.

1. **Film.** Most people have seen an x-ray picture. The picture is the result of radiation hitting a photographic film. The more radiation exposure, the more blackening of the film. Film-badges are often used by people working with radiation in hospitals or in research, keeping track of how much radiation exposure the workers have received.

2. **TLD.** These initials stand for "Thermo Luminescence Dosimetry". A crystal such as LiF containing Mn as an impurity is used. The impurity causes traps in the crystalline lattice where, following irradiation, electrons are held. When the crystal is warmed, the trapped electrons are released and light is emitted, the amount of light being related to the dose of radiation received.

3. **Ionization.** Radiation results in the formation of positive and negative ions in a gas. The number of ions formed can be detected. The famous Geiger–Mueller tube, commonly called a Geiger counter, is designed to measure the electrical response produced by the newly formed ions.

4. **Scintillation.** A number of compounds will emit light when exposed to radiation. The intensity of the emitted light depends on the radiation exposure and the light intensity is easily measured.

5. **Semiconductors.** Radiation produces an electric current in semiconductors that can be measured.

6. **Free radicals.** Radiation produces a class of chemical species known as free radicals. Free radicals by definition contain an unpaired electron and, although they are very reactive, they can be trapped in some solid materials. The number of trapped free radicals is a measure of the radiation dose.

7. **Redox products.** Radiation either reduces (by electron addition) or oxidizes (by electron abstraction) the absorbing molecules. Although these changes are intially in the form of unstable free radicals, chemical reactions occur that ultimately result in stable reduction and oxidation products.

*On the following pages we give some more informa-
tion about the different types of equipment for those
interested. Much more information can be found in
textbooks on radiation dosimetry.
(Those not interested in measurement devices can skip
to page 54.)*

Film

This simple device uses photographic film. Ionizing radiation interacts with film
pretty much like ordinary light. The sensitive compound is silver bromide. It is
split by radiation into atomic silver and bromine, leading to the formation of
metallic silver particles. Since the silver particles are black, the darker the film
the higher the dose.

Workers who are potentially exposed to radiation often use film as a personal
dosimeter (an example is given in Figure 6.1). It is worn on work clothes
(sometimes on the lapel or belt). The film is contained in a plastic holder that
has small absorption plates of lead, tin, cadmium and plastic. In addition, there
is an open window that makes it possible for weaker radiations to reach the
film. The blackening behind the different plates depends on the energy and type
of the radiation.

If the radiation contains neutrons of low energy, called thermal neutrons, the film
behind the cadmium plate will show some extra blackening because of reactions
between neutrons and cadmium.

*Figure 6.1. A picture of a film dosimeter. To the right, the plastic holder is
opened to show the construction. In addition to an open window you can see
several areas covered with absorption plates of different types.*

Thermoluminescence dosimeter (TLD)

Following a radiation exposure, electrons trapped in the crystalline lattice, when heated, receive enough energy to escape from the trap and fall to the ground state emitting light photons. Since warming is a requirement, the technique is called thermoluminescence. The intensity of the luminescence is a measure of the dose.

Small crystals of LiF (lithium fluoride) are the most common TLD dosimeters since they have the same absorption properties as soft tissue. The amount of light emitted at 200°C due to the radiation is proportional to the dose in soft tissue. It may be used as a personal dosimeter for β-and γ-radiation because it is independent of the energy of the radiation. Lithium has two isotopes, Li-6 and Li-7. Li-6 is also sensitive to neutrons. Consequently, if there is a combination of neutron and γ-radiation, the light emitted from a LiF crystal with Li-7 is a measure of the γ-radiation, whereas the light emitted using Li-6 yields the total dose from both neutrons and γ-radiation. In the case of neutron detection, the efficiency of the detector depends on the energy of the neutrons. Because the interaction of neutrons with any element is highly dependent on energy, making a dosimeter independent of the energy of neutrons is very difficult.

Detectors based on ionization

An ionization chamber consists of a gas volume in an electric field between two electrodes. Radiation entering this volume results in the formation of ions. The positive ions will be attracted to the negative electrode, and negative ions will be attracted to the positive electrode. Ions with high enough energies may ionize even more molecules on their way to an electrode. This means that when the voltage across the electrodes increases, the number of ions increases. For a certain voltage (the proportional region), the number of ions at the electrode is proportional to the radiation energy deposited in the gas volume, resulting in a qualitative measure of the radiation energy.

If a very high voltage is used (called the Geiger–Mueller region), each ionization yields a cascade of ions that results in a pulse. Regardless of the energy of the radiation, the same size pulse is formed.

The following instruments use the ionization process:

Proportional counter. For this instrument the voltage across the electrodes is adjusted to let the number of ions that reach the electrodes be equal or proportional to the number of ions induced by the radiation. Thus, the counter can distinguish between α- and β-particles and even determine their energy.

Geiger–Mueller counter. This famous instrument, shown in Figure 6.2, is named after the two physicists who invented the counter in 1928. It consists of a gas volume with two electrodes that have a high voltage between them. Very often the detector element is cylindrical in shape with the cylinder wall serving as the negatively charged (ground) electrode and a thin metal rod running along the middle axis serving as the positively charged electrode.

Ionizing radiation passing through the gas volume produces ions in the gas. The voltage is high enough for each electron attracted to the central electrode to make a cascade of new ions. This results in a pulse which is detected by a counter system and is also sent to a speaker which produces an audible click.

The counter can be used as a warning instrument. It does not, however, yield information about the type of radiation or its energy.

The requirement that the radiation should reach the sensitive gas volume may be difficult for α-particles. The G-M counter is well suited to localize β-, γ-emitting radioactivity in connection with accidents.

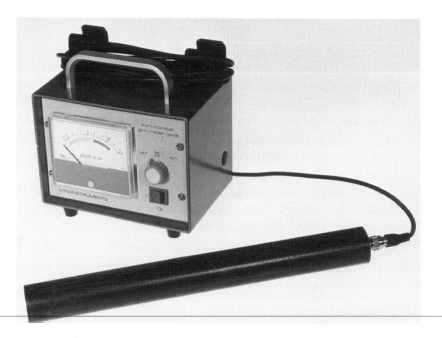

Figure 6.2. A Geiger–Mueller counter. The sensitive volume is a tube which is connected to a pulse counter. The counter yields information about the radiation intensity.

Scintillation Counter

The scintillation counter is based on the principle that light is emitted when a scintillator is exposed to radiation. Several organic compounds act as scintillators, such as benzene and anthracene. However, the most commonly used scintillator is a crystal of sodium iodide (NaI). The light pulse produced is recorded by a photomultiplier tube.

Single crystals of NaI can be made with volumes of several liters. Hospitals all over the world have gamma-cameras which use NaI crystals in the form of large plates with diameters of more than 40 cm.

The light emitted when the crystal is irradiated is proportional to the γ-energy deposited. Consequently these counters are suited to measure the energy of γ-radiation and, therefore, can be used to identify γ-emitting isotopes (see Figures 6.3 and 6.4).

Semiconductor Counters

Transistors and other components in electronic equipment are made of semiconducting materials. When electrons are released in these semiconductors, the current can be measured with great accuracy.

Solar cells are made of thin silicon crystals. They give rise to electric current when hit by solar light. In a similar way a current can be induced by ionizing radiation. A large, clean and almost perfect semiconductor is ideal as a counter for radioactivity. The released electric charge is closely related to the radiation energy and the charge (or charge pulses) is detected by the electrical change in the semiconductor. These counters are employed to measure the energy of the radiation.

The crystals are made of silicon or germanium. However, it is difficult to make large crystals with sufficient purity. The semiconductor counters have, therefore, low efficiency, but they do give a very precise measure of the energy. In order to achieve maximum efficiency the counters must operate at the very low temperatures of liquid nitrogen (–196°C).

Free Radical Dosimetry

Among the highly reactive products formed in materials by radiation are free radicals. By definition, a free radical is a molecule that has an unpaired electron. Some examples are the hydroxyl radical (OH), the methyl radical (CH_3), and nitrous oxid (NO). Radicals are very unstable in aqueous systems such as in a living cell, but their lifetimes can be quite long in solid materials. The radicals formed in crystalline alanine (a simple amino acid) have become very popular in dosimetry. The number of radicals formed, which is proportional to the dose, is observed with a technique called electron spin resonance (see Chapter 12). The advantage of this dosimeter is its small size and its electron density is comparable to soft tissue. Furthermore, this type of dosimeter can be used in radiation accidents where other dosimeters often fail. An example of this is given in Chapter 10.

Chemical Dosimetry

Ionizing radiation also produces chemical changes which can be observed and measured. The radiation dose can be determined by observing the amount of a chemical product which is formed. A well-known chemical reaction is based on the oxidation of Fe^{++} to Fe^{+++}. It was the Danish chemist H. Fricke, who introduced this reaction for dosimetry in 1929. He used iron sulfate in a solution of sulfuric acid and observed the amount of Fe^{+++} ions formed. Chemical dosimetry is used mainly in radiation chemistry studies.

The Energy of the Radiation

It is very important to have sufficient information about the energy of the radiation. With information about the energy and the type of particle observed it is possible to identify a radioactive isotope. This may have great importance with regard to the storage and handling of radioactive sources.

Both semiconductor counters and scintillation counters produce pulses which are a measure of the radiation energy deposited in the sensitive volume. The number of pulses is a measure of the activity. Before they are measured, the pulses are electronically amplified and filtered and grouped according to their size. They are then converted to numbers in an analog to digital converter (ADC). A computer sorts out the numbers and presents the results.

Figure 6.3 shows the results for Cs-137 observed with a sodium-iodide scintillation counter and a germanium semiconductor counter. The radiation energy is given along the abscissa and the count number is given along the vertical axis. The curve attained is called a γ-spectrum. Cs-137 emits both a β-particle and γ-ray. It is the γ-photon with an energy of 662 keV that is observed. The figure demonstrates how the two counters react when hit by the radiation.

Energy is absorbed either completely (the photoelectric effect) or partly in a Compton process. The peak in the spectrum has an energy of 662 keV and is called the *"photopeak"*, whereas the partly absorbed energy is seen in the left part of the spectrum, the Compton part. The photopeak is used for identification of the isotope.

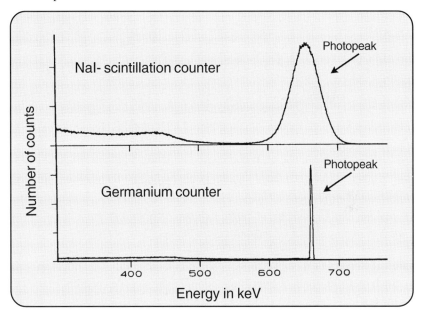

Figure 6.3. The γ-ray spectrum for Cs-137 measured with a sodium iodide scintillation crystal and a germanium semiconductor counter. The γ-radiation has a peak at 662 keV (the photopeak). This is important for identifying the different isotopes. (Courtesy of Finn Ingebretsen, Inst. of Physics, Univ. of Oslo)

The curve is usually called the response function of the counter. If a radioactive source emits several photons with different energies the spectrum will consist of several photopeaks.

It is seen that a scintillation counter yields much broader lines compared to the germanium counter. Generally, it is easier to identify different isotopes with a counter that has good energy resolution, such as the semiconductor counter.

An example with several isotopes is given in Figure 6.4. The fallout from the Chernobyl accident was observed two weeks later and about 2,000 km away in Norway. The γ-spectrum shows that a number of isotopes were released. The isotopes that emit γ-radiation can be identified.

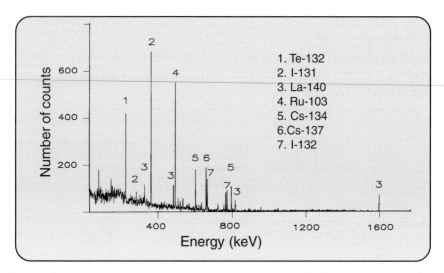

Figure 6.4. A γ-ray spectrum (germanium counter) which makes it possible to identify seven different isotopes. The sample, consisting of grass from the lawn outside the University of Oslo, exhibits the fallout from the Chernobyl accident. (Courtesy of Finn Ingebretsen, Inst. of Physics, Univ. of Oslo)

Chemical Separation can be used to identify isotopes that do not emit γ-radiation. Such methods made it possible to identify the isotopes present in the fallout from the nuclear tests in the 1960s. For example, it was found that the amount of Sr-90 (a pure β-emitter) was equal to the amount of Cs-137.

Dose Measurements

The strength of a radioactive source (in Bq) and the energy of the emission (in eV) can be measured. This is, however, not a dose measurement.

The radiation dose is the energy deposited in the irradiated compound. If the radiation hits a human being, the dose is defined as the energy deposited in the human body. The amount of energy deposited is almost always different from the amount of energy coming from the source. Deposited energy determines dose.

Counters observe particles or photons sequentially. In dose measurements, the concern is not with the individual particles or photons but with the total energy absorbed in the exposed materials (e.g., tissue). It is difficult to observe energy absorption in tissue. Two of the problems are:

1. An exposure to one roentgen (1.0 R) of x- or γ-radiation results in a radiation dose to soft tissue of approximately 9.3 mGy. The precision can be no better since the roentgen unit is based on the radiation absorption in air, whereas doses to a biological system (soft tissue or bone) are based on the energy absorbed in that system.

   The absorption increases with the electron density of the exposed material and is therefore larger in bone compared with soft tissue. Furthermore, the energy absorption increases with decreasing radiation energy. Since these properties *are not* the same for air, soft tissue and bone, the doses delivered by a 1 R exposure are different.

2. When the radiation strikes a body, the dose changes with depth (i.e., the distance the radiation traverses in the body). This is illustrated in Figure 6.5 for different types of x-rays as well as a beam of charged particles (C-12 ions).

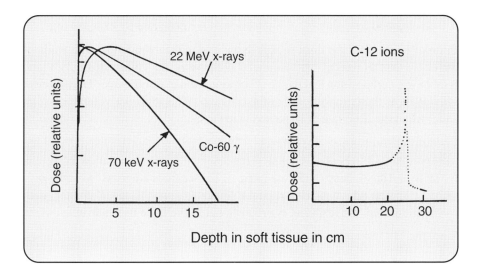

Figure 6.5. Depth dose curves for soft tissue. The dose is measured from the surface of the skin. On the left are data for x- and γ-rays as indicated. To the right is shown the dose curve for high energy charged particles. In this example carbon atoms, with all 6 orbital electrons stripped away, were used. The energy of the carbon ions when they hit the soft tissue is 5,688 MeV. (Produced at the Bevalac that used to operate at the Lawrence Berkeley Laboratory in Berkeley, California, USA.)

In order to use radiation for cancer treatment it is important to have knowledge of these depth dose curves. As you can see from Figure 6.5, the region for maximum dose can be changed by changing the x-ray energy. For tumors positioned deeper than 6 cm into the tissue, x-rays with an energy of more than 20 MeV should be used. The goal is to give a high killing dose to the tumor while minimizing the dose to the surrounding healthy tissue.

As you can see, the charged particle radiation has a striking depth dose relationship (Figure 6.5). The dose peaks at the end of the track (called the Bragg peak). The Bragg peak occurs at a depth which depends on the energy of the particles. This type of radiation has been used for cancer treatment. It requires large accelerators and it is, therefore, quite expensive. The depth dose curve shown in Figure 6.5 was obtained with a research accelerator at the Lawrence Berkeley Laboratory in the 1960s; it is no longer in operation.

The "**dose rate**" is usually given when describing the intensity of the radiation being absorbed by the target, i.e. the radiation dose per unit time. The total dose is then obtained by a simple multiplication of the dose rate by exposed time.

In order to measure the dose to soft tissue exposed to γ-radiation, the walls of the instrument as well as the compound in the sensitive volume (usually a gas) must have a composition that has an absorption similar to soft tissue. The observed ion current is then proportional to the dose rate. The sensitive volume must be small when the dose rate is large and vice versa.

Thermoluminescence dosemeters (TLD) are frequently used for dose measurements. TLDs are well suited for different types of radiation as well as for large and small doses and dose rates. If LiF-crystals are used, x- and γ-rays yield a response which is proportional to the dose to soft tissue. If crystals with calcium are used, the response is proportional to the bone tissue dose.

Since the TLD crystals are small, they are well-suited for measurement of doses to patients. They can be placed on and in the body.

In this section we summarize the guidelines for radiation protection. Most or all countries follow the recommendations from ICRP. The recommendations have been changed several times during the years when more information became available. The present guidelines are based on the linear dose effect curve with no threshold. We discuss this in more detail in Chapter 11.

Radiation Protection

Before considering the biological effects of radiation, the regulations for radiation protection will be discussed briefly.

Radiation is used in medicine, in research and in industry. These applications of radiation are useful to society. However, radiation can also have detrimental effects. It was important to establish rules and regulations governing these uses to balance the positive uses against the negative effects. In other words, there must be a balance between benefit and risk. Thus, in order to use a radioactive source in a hospital, the benefit to the patient (and society) must be larger than the damage to those who make, transport and use the source.

Dose limits have been established for groups who use radiation in their work as well as for the population at large. The values are set mainly for *radiation workers*. For the rest of the population, the dose limits are approximately 10% of those set for radiation workers.

The rules used in most countries have been worked out by the *International Commission on Radiological Protection* (ICRP).

The ICRP has adopted the following principles for the use of radiation:

> The application of radiation should be useful.

This means that all activity involving radiation will only be allowed when the benefits exceed the risk. This principle is widely applied today.

> The radiation dose should be as low as reasonably achievable (the ALARA-principle).

This principle means that large resources should not be used to reduce doses that are already small, but radiation doses should be reduced whenever it is reasonable to do so. Let us give an example:

In previous years watches were used where the numbers and pointers were painted with radium and a fluorescent compound. The radium used emitted strong γ-radiation and yielded unacceptable doses to the skin and the gonads to those wearing the watches. Today, these radium watches are no longer used and instead of radium, β-emitters are used. The reason, of course, is that the β-particles will not be able to escape from the watch dials. Consequently, the radiation dose to the wearer is rendered insignificant.

Dose Limit Values

Doses yielding a personal risk to workers and the public rarely occur. The maximum allowed by law is low. In the rules for radiation workers, the leading principle is that the risk for radiation damage should not be larger than the acceptable risk of other occupations.

The ICRP suggests the following values:

Radiation workers. The limit is 20 mSv per year, (averaged over 5 years). The doses should not exceed 50 mSv for one particular year.

The population. Radiation, except for background radiation, should not exceed 1 mSv per year.

Pregnant women. Particular control is applied to radiation workers who are pregnant. If and when pregnancy is ascertained, the dose to the fetus should be kept below 1 mSv for the rest of the pregnancy. The reasons for this are explained in Chapter 12.

International Committees and Societies

Several international committees consider the effects of radiation on humans. For radiation protection the **ICRP** is the leading body, whereas **UNSCEAR** reviews epidemiological data. The majority of scientists working on the biological effects of radiation are gathered in organizations such as the ***Radiation Research Society (RRS)***. Scientists present the findings of their research at conferences and in journals.

Rolf M. Sievert

Courtesy of Swedish Radiation Protection Institute

ICRP – The *International Commission on Radiological Protection* was established in combination with the second international congress in radiology in Stockholm, Sweden, in 1928. Rolf M. Sievert was one of the founders and later became chairman. The ICRP makes suggestions and rules for radiation protection, which are followed in most countries that have established a national protection board. The ICRP is, and should be, conservative with regard to its recommendations. The ICRP uses the principle of linear dose–effect-curves with no threshold doses (see Chapter 11) to estimate the health effects from small radiation doses.

UNSCEAR – *The United Nations Subcommittee on the Effects of Atomic Radiation* (UNSCEAR) was established in 1955 with R.M. Sievert as the first chairman. The purpose is to collect and judge research data from all over the world concerning radiation and health. The committee reports its work to the UN general assembly. The reports are usually concentrated on particular subjects. Thus, in 1986, the genetic effects of radiation were discussed; in 1988, the subject was radiation from natural sources as well as the Chernobyl accident, and in 1993, it was a more than 1000-page report on doses from natural and artificial sources. In 1994, one of the subjects was a strongly debated issue, namely whether small doses of radiation may have a beneficial health effect. In particular, the important results that indicate that small doses may stimulate the immune system (so-called adaptive effect) were discussed (see Chapters 11 and 12).

Gioacchino Failla

RRS – The *Radiation Research Society* was incorporated in 1952 under the leadership of Gioacchino Failla. It is an international society that has three objectives: to encourage in the broadest manner the advancement of radiation research in all areas of natural sciences; to facilitate cooperative research between the disciplines of physics, chemistry, biology and medicine in the study of the properties and effects of radiation; to promote dissemination of knowledge in these and related fields through publications, meetings and educational symposia.

Courtesy of Radiation Research Society

Radioactivity in your body

There are natural isotopes (primarily K-40) in our bodies that deliver an annual radiation dose of 0.3 to 0.4 mSv (see next Chapter). The nuclear tests in the atmosphere in the 1960s and the Chernobyl accident have resulted in fallout and pollution from a number of radioactive isotopes. Some people are concerned that new accidents and leakage from nuclear storage centers may make the pollution worse. Some isotopes will eventually come into the food chain and into our bodies. As long as the amount of radioactive isotopes is small, the radiation doses will be negligible. However, the following question arises. *Where should the limit for artificial radioactive isotopes in our food be set?*

Radioactive isotopes with short half-lives such as I-131 yield a temporary dose problem. The iodine isotope is important since iodine concentrates in the thyroid gland. This effect appears to be the most important health concern of the Chernobyl accident.

The isotope Cs-137, with a physical half-life of 30 years, has also caused considerable concern after the Chernobyl accident. This isotope was also a major fallout product after the nuclear tests in the atmosphere in the 1960s. Shortly after the Chernobyl accident, several European countries decided to set an upper threshold level with regard to radio-activity in food products.

Thus, if the level was above the set limit, the products were not allowed to be sold. In certain countries 600 Bq per kilo was used, which meant that large amounts of meat from sheep and reindeer could not be used.

Is it dangerous to eat food with a Cs-137 content of 1000 Bq/kg?

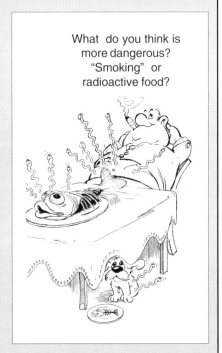

What do you think is more dangerous? "Smoking" or radioactive food?

This book provides you with the knowledge needed to answer that question. To give you an idea, consider the following. A dinner with meat containing 1,000 Bq/kg yields a radiation dose of about 0.003 mSv. Ten such dinners are equivalent to the extra dose attained on a commercial flight between Europe and the US! If you have such dinners every day throughout the year, the extra dose is still smaller than that received from natural radiation.

Chapter 7

Natural Radiation

Introduction

People live surrounded by natural radioactive sources. There are radioactive isotopes in our bodies, houses, air, water and in the ground. In this Chapter, natural radiation will be discussed in more detail, in particular how much natural radiation people are exposed to every day.

Common sources of radiation are presented in Figure 7.1. The annual doses vary from one area to another but are more or less equal to those shown in the figure.

Cosmic Radiation

Annual dose = 0.3 to 0.6 mSv

The atmosphere is continuously exposed to particles from outer space. A stream of particles consisting of protons (about 85%), α-particles (about 13%), and a small fraction of larger particles hit the outer atmosphere. Some of the particles have very large energies when hitting the atmosphere (up to 10^{14} MeV). When the particles interact with the atoms in the atmosphere their energy gradually decreases and a number of *new* high energy particles are formed.

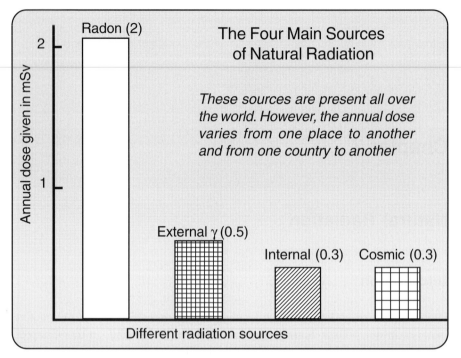

Figure 7.1. Natural radiation sources and their annual doses. The particular dose levels given are for Scandinavia. The annual doses are given in mSv.

When cosmic radiation reaches both the inner atmosphere and the ground, it is accompanied by γ-radiation, neutrons and various other small particles. In addition, some new radioactive isotopes are formed. The more important of these, *cosmogenic radionuclides,* are H-3 (tritium), Be-7, C-14 and Na-22.

In addition to the radiation from space, which is rather constant, cosmic radiation contains particles and electromagnetic radiation from the sun. The radiation from the sun is variable due to sunspot activity (which has a period of about 11 years). The energies of these solar particles are small, contributing only a tiny fraction of the total dose from cosmic radiation.

The contribution of neutrons that accompany cosmic radiation increases with height above sea level. This influences the dose since neutrons have a radiation weighting factor of 5 to 20, resulting in an annual dose at sea level of approximately 0.3 mSv. For those living 1,500–2,000 meters (5,000–6,000 feet) above sea level, the annual dose is approximately doubled. If you climb up to 5,000 meters (15,000 feet) the radiation is ten times that at sea level, mainly due to the increased neutron contribution.

The amount of cosmic radiation varies somewhat with latitude. Particles from space have an electric charge and the motion of these charged particles is influenced by the magnetic field around the Earth. For tropical areas this inter-action is largest, deflecting the particles while they are still high up in the atmosphere. In polar regions, the path of the particles more closely follows the magnetic field lines, allowing them to penetrate deeper into the atmosphere, which results in larger doses near the poles of the Earth. Another effect, due to the excitations of the atmospheric molecules, is the beautiful northern and southern lights (*aurora borealis* and *aurora australis*).

Air travel

The variation in cosmic radiation with the height above sea level can easily be observed when flying. This means that air crews with a large number of flying hours receive an extra annual dose. Scientists in Germany supported by the "Gesellschaft für Strahlenforschung" placed dosimeters inside airplanes in order to measure the amount of γ- and neutron radiation. From their observations it was calculated that, for an annual flying time of 600 hours at about 10,000 meters (30,000 feet), the extra dose is 3 mSv. For a height of 18,000 meters (55,000 feet), the dose rate is about 0.15 mSv per hour. If, for example, you use the above data for your own air travels, you will find that a trip between Europe and US yields an extra dose of about 0.03 to 0.04 mSv. The degree to which these small extra doses might increase risk is the focus of Chapter 11.

Natural Gamma Radiation

Annual dose = 0.5 mSv

Gamma-radiation, emitted by radioactive isotopes in building materials and the ground, produces an annual dose of about 0.5 mSv. In certain areas the γ-radiation can be much larger, particularly in areas where the ground contains thorium and radium. Let us consider, in more detail, the sources of γ-radiation.

Natural radioactive sources

A long time ago, when the earth was created, a number of radioactive elements were formed. Some of these isotopes have very long half-lives, billions of

Air Travel and Radiation

The intensity of cosmic radiation increases with altitude. A considerable fraction of the high energy radiation will penetrate airplanes, giving passengers and crew an extra dose during air travel. This extra dose depends on the altitude and the length of the trip. In order to show this, the measured results for a short trip of only 40 minutes reaching an altitude of 8500 meters (27,900 feet) are presented.

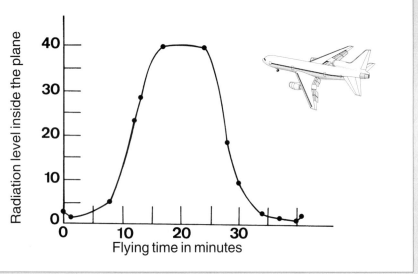

Courtesy of Hans Bjerke, Norwegian Radiation Protection Authority

When a plane is on the ground, the radiation level is set equal to 1. The value of 1 represents the sum of cosmic radiation reaching ground level and γ-radiation emanating from natural sources at ground level. As the plane reaches the altitude of 8,500 meters (27,900 ft), cosmic radiation increases to approximately 40 times that observed at ground level.

As you can see, the radiation dose was slightly smaller just after take-off and just before landing. The reason for this is that, in both occasions, the plane was over water, which contains very low levels of γ-emitters.

The radiation dose due to air travel is in addition to the doses from other natural sources. The extra dose accumulated annually by air crew members while they are in the air (2 to 4 mSv) is comparable to the equivalent dose accumulated on the ground due to natural sources.

A single flight from the US to Europe yields the same extra dose as 10 dinners with meat containing Cs-137 to a level of 1000 Bq/kg. After the Chernobyl accident, the maximum level for food products was set to 600 Bq/kg in most European countries.

years, and so they persist today. Since radioactive isotopes are unstable, a dis-integration eventually occurs, resulting in a different type of atom. This continues until a stable (non-radioactive) isotope is formed. Because most of the natural radioactive isotopes are heavy (found in the fifth row or higher in the periodic table), more than one disintegration is necessary before a stable atom is reached. This radioactive series is called a *radioactive family*. It is not unusual for the number of disintegrations comprising a series to be around 11 to 14. The two most important radioactive series today are the *uranium–radium-series* and the *thorium-series*. There are two other series, but they have almost disappeared from the Earth because they have much shorter half-lives. The four possible series are given in Table 7.1.

Table 7.1. The radioactive series

Name	Start	End	Half-life
Uranium-radium	U-238	Pb-206	4.47×10^9 year
Neptunium	Np-237	Bi-209	2.14×10^6 year
Uranium-actinium	U-235	Pb-207	7.038×10^8 year
Thorium	Th-232	Pb-208	1.405×10^{10} year

The first radioisotope in the thorium-series has a half-life of $1.4 \cdot 10^{10}$ years. This is why there is still a lot of thorium in the ground. On the other hand, neptunium with its short half-life has disappeared. However, it can be produced in the laboratory. Thus, all four radioactive series are well understood. The starting element in the uranium–actinium series has a half-life of $7.038 \cdot 10^8$ years. This is short compared to the age of the Earth (5 billion years or about 7 half-lives), and therefore the content of U-235 is only 0.71 % of the uranium isotopes.

In addition to these radioactive series, there are a number of other radioactive isotopes. The most important is K-40 with a half-life of 1.27 billion years. This half-life is shorter than the age of the Earth and only a few percent of the original K-40 remains today. The half-lives given in Table 7.1 indicate that the natural radioactivity has decreased considerably since the formation of the Earth. This is of interest when speculating about the origin of life and discussing the possible effects of radiation when the Earth was younger.

Distribution of natural radioactivity

Natural radioactivity varies from place to place. With regard to doses from external γ-radiation, the most significant contributions come from the elements in the uranium–radium-series, the thorium-series and K-40.

In order to give a quantitative measure for the presence of radioactive elements, the number of becquerel per kilogram (Bq/kg) can be obtained for different types of rock and soil. Of course, this varies from place to place.

Table 7.2. The concentration of isotopes (given in Bq/kg) in some species of rock and soil

Species of rock/soil	Ra-226	Th-232	K-40
Granite	20–120	20–80	600–1800
Thorium and uranium rich granite	100–500	40–350	1200–1800
Gneiss	20–120	20–80	600–1800
Sandstone	5–60	4–40	300–1500
Limestone	5–20	1–10	30–150
Slate	10–120	8–60	600–1800
Shale (from cambrium)	120–600	8–40	1000–1800
Shale (lower ordovicium)	600–4500	8–40	1000–1800
Shale rich soil	100–1000	20–80	600–1000
Moraine soil	20–80	20–80	900–1300
Clay	20–120	25–80	600–1300

Courtesy of Anders Storruste, Inst. of Physics, Univ. of Oslo

Table 7.2 gives the concentrations of Ra-226, Th-232 and K-40 in different species of rock found in Scandinavia. It appears that certain types of shale exhibit concentrations of Ra-226 up to 4,500 Bq/kg. The concentration of Th-232 also varies considerably from one mineral to another. In certain areas of the world, such as India, Brazil and Iran, the thorium concentration in the soil can be 10 to 100 times above the average.

Potassium is everywhere. It is in soil, plants, animals, and humans. The element potassium makes up 2.4 percent by weight of all elements. But, the abundance of the radioactive isotope K-40 is only 0.0118%.

Notice in Table 7.2 that the variation of K-40 between different types of rock is much smaller than that for radium and thorium. K-40 emits both a β-particle and a γ-ray.

Each isotope contributes a different amount to the total external γ-dose: approximately 40% from K-40, 40% from Th-232 and about 20% from Ra-226.

When inside houses and buildings, people are shielded from much of the radiation coming from outside. But since the building materials also contain radioactive elements, the dose generally does not go down. More often it goes up. Since the concentration of radioactive isotopes varies from one region to another, it is also true that the radioactivity in building materials, such as concrete, depends on where it is made.

In houses made of wood, the γ-radiation in the ground floor is approximately the same as that outdoors. This is because most of the radiation comes from the masonry materials in the cellar and the ground just outside. Wood materials contain less radioactivity than rocks and soil and the radiation level decreases as you go to higher floors. In houses made of concrete, the indoor radiation level will be like that of outdoors if the concrete is made from materials found locally. But it is not unusual for the concrete or cinder blocks to contribute to an increase in the radiation level over that found outdoors.

In Sweden, a number of houses were built using uranium rich shale. Because these houses have a radiation level that is unacceptable, construction using this type of building material was stopped in 1979. Houses made of red brick very often have a high radiation level, mainly because of the content of K-40. Brick houses are more common in cities than in the country; consequently, people living in cities tend to be exposed to more radiation than those living in the countryside.

Examples of the radiation level inside different types of houses in the same area are given in Figure 7.2. These data, collected by graduate students, are for more than 2000 houses in Southern Norway. For different types of houses, the radiation level may vary by a factor 2. The lowest levels were found for wood houses. Living in an average wood house for one year would result in a whole body dose of 0.87 mGy (for this type of radiation the weighting factor is 1 and the biological effective dose is the same, 0.87 mSv) from external γ-radiation and cosmic radiation.

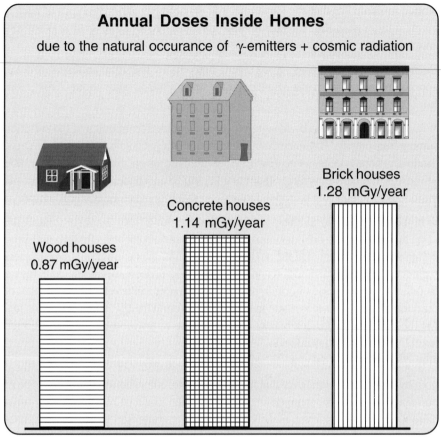

Figure 7.2. Here is an example that shows that the annual dose depends on the house a person lives in. Approximately 2000 dwellings are included in this study. The value given for each type of dwelling is a combination of cosmic radiation and γ-radiation from sources external to the human body. The dose rate is given in mGy per year. Since this is mainly low LET-radiation, the dose rate would be the same in mSv. The houses in this examination are in Southern Norway, but similar results are likely in other areas.

It appears from the figure that, if you move from one house to another, you will also change your radiation milieu, your background dose either increasing or decreasing. Consequently, if small environmental doses, either natural or artificial, are of some concern, it would be of interest to have information about the sources that contribute to annual dose. The external γ-radiation dose (coming from sources external to the human body) varies within large limits from place to place, and even from one house to another house in the same area.

Internal Radioactivity (sources inside the body)

Annual dose = 0.3 – 0.4 mSv

The food we eat contains radioactive isotopes and our bodies will therefore contain small amounts of radioactivity. The most important isotope is K-40. The daily consumption of potassium is approximately 2.5 gram. From this you can calculate that each day you eat about 75 Bq of K-40. Potassium is present in all cells making up soft tissue. The potassium content per kilogram body weight will vary according to sex and age (see Figure 7.3). The dose due to K-40 will of course also vary in a similar way. Muscular young men receive a larger dose than older persons and men receive a larger dose than women.

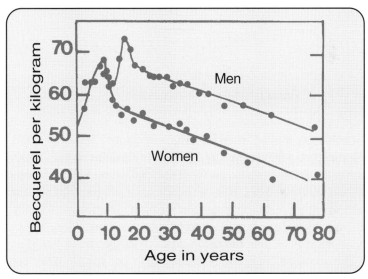

Figure 7.3. The concentration of potassium (in becquerel per kilogram from K-40) varies with age and sex. The abundance of the K-40 isotope is 0.0118%.

Two other naturally occurring radioactive isotopes of concern are C-14 and polonium (Po-210).

C-14 is formed in the atmosphere when neutrons react with nitrogen. C-14 is taken up by plants in the form of carbon dioxide, CO_2.

Polonium is one of the decay products in the uranium-series and results from the decay of the noble gas radon (see Figure 2.3). After emitting two β-particles,

Po-210 is formed. This isotope is an α-particle emitter and, if ingested or inhaled, it delivers a relatively large local dose. When taken up by the body, 10% of the isotope is deposited in the liver, 10% in the kidneys, and 10% in the spleen. The remaining 70% is uniformly distributed throughout the body.

The amount of polonium in food varies. It is particularly high in reindeer and caribou meat because it concentrates in lichen, an important food source for these animals. Some people living in Lappland eat a lot of reindeer meat and, consequently, have Po-210 concentrations much higher than average.

The tobacco plant also takes up Po-210; thus, smokers get an extra radiation dose to their lungs.

All of the different natural isotopes in your body yield an annual dose of about 0.3 to 0.4 mSv.

Radon

Annual dose = 1 – 3 mSv

Radon is a noble gas formed when radium disintegrates. Since radon is a gas, it readily becomes part of the atmosphere. The half-life is rather short (3.82 days) and its four subsequent decay products (also with short half-lives) add large natural doses to the public environment. Some areas are high in radon because local concentrations of uranium are high. Radon measurements in dwellings, together with mathematical models, have been used to estimate the annual dose to the public. The calculations indicate that the radon doses to the people in Scandinavia, as well as to the average US citizen, is about 2 mSv per year. In other countries, like Denmark and Iceland, the doses are much smaller. In mines and other structures in the mountains (hydroelectric power stations and military installations), the radon concentrations may be large. People working at such places receive larger doses.

Several possibilities exist for the release of radon into houses. The main sources are the rock or soil on which the house is built, as well as the water supply. The rock formations under a house always contain some radium and the radon gas can penetrate into the house through cracks in the floor and walls of the basement. The water supply from wells, in particular in regions with radium-rich granite, contains high radon concentrations. When the water is the carrier, radon gas is readily released.

The drawing in Figure 7.4 shows the main sources for indoor radon. In some areas, the ground and building materials are the most important sources, whereas in other areas the water supply is more important.

Rn-222 is a noble gas and is an α-particle emitter. The four decay products: Po-218, Pb-214, Bi-214 and Po-214 are usually called *radon decay products* or *radon daughters*. They are metals and have half-lives of only minutes. When inhaled, they may be deposited in the bronchia. It is, therefore, the lungs and the bronchia that receive radon doses. The long-lived products that follow the radon daughters yield small doses to other parts of the body.

The highest radon concentrations can be found in areas with shale. The magnitude of the radon content within houses is also determined by construction and architecture. Consequently, the radon concentration in two adjacent houses may be quite different.

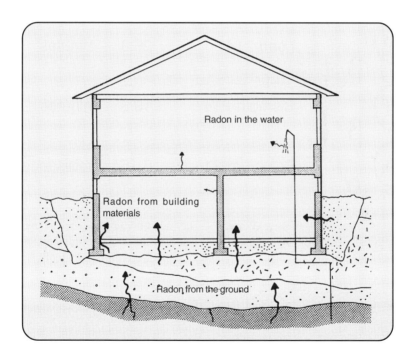

Figure 7.4. This drawing illustrates how radon enters our houses. Radon is a gas originating from radium in soil, rock or water. It also comes from small amounts of radium in the building materials. (Courtesy of Terje Strand, Norwegian Protection Authority)

The Radon Family

The isotope Rn-222 and the four following decay products in the uranium–radium series is called the *radon family*. Radon is a gas, but the 4 decay products (polonium-218, lead-214, bismuth-214 and polonium-214) are all metals. The table below summarizes the data.

		Half-life	
Radon-222		3.82	days
Polonium-218		3.05	minutes
Lead-214		26.8	minutes
Bismuth-214		19.8	minutes
Polonium-214		0.164	milliseconds

Ra-222, Po-218 and Po-214 are α-emitters. The energy of the particles range from 5.49 MeV to 7.69 MeV.

Pb-214 and Bi-214 are β-emitters with an average energy of 0.29 MeV and 0.65 MeV, respectively. The isotopes also emit γ-radiation. Thus, Pb-214 has an average γ-energy of 0.25 MeV and Bi-214 has an average γ-energy of 1.46 MeV.

The radon decay products can attach to dust particles in the air. When inhaled, some of this radioactive dust collects inside the lungs.

The α-particles from Po-218 and Po-214 yield the main fraction of the radiation dose. The dose calculations are difficult. They are based on models that have many uncertainties, such as the radiation weighting factor to be used in order to calculate the dose in Sv (a factor of 20 is used).

The radon concentration is measured in becquerel per cubic meter (Bq/m³). For ordinary dwellings the concentration is usually below 100 Bq/m³. It is not constant and varies through the year. The World Health Organization (WHO) recommends remediation when the concentration exceeds 800 Bq/m³. For concentrations between 200 and 800, simple and cheap remediation may be considered. There are cases where the concentrations inside the houses have reached 10,000 Bq/m³ or more.

Remediation involves methods that prevent radon from seeping into the house from the ground and through cracks in the concrete walls. Another remediation technique is to "suck out" the radon-rich air from underneath the cellar floor. A tube and a fan can pump the radon to the outside as shown in Figure 7.5. This solution has proved to be efficient for a number of houses high in radon. In order to build a new house in an area with high radon concentration the problem can be reduced significantly if diffusion-dense materials are placed underneath the foundation wall.

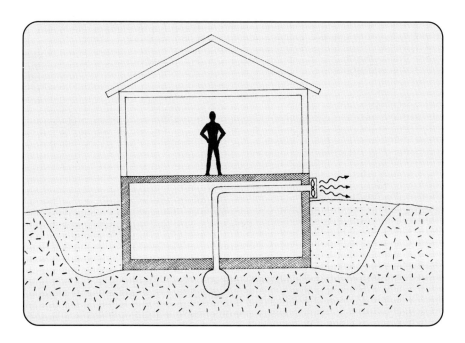

Figure 7.5. Remediation in a house with a large concentration of radon. A simple fan can "pump" the radon out of the ground and prevent it from seeping into the house. (Courtesy of Terje Strand, Norwegian Protection Authority)

Measurement of radon

The radon concentration in a house varies considerably with the time of the year and the ventilation. One example is shown in Figure 7.6.

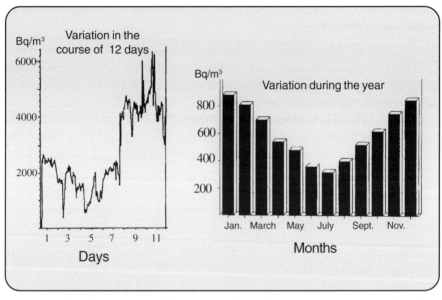

Figure 7.6. The Figure exhibits examples of radon variation in a dwelling over a short-time (12 days) and a long-time (12 months). (Courtesy of Terje Strand, Norwegian Protection Authority)

As can be seen in Figure 7.6, there is an annual variation with the largest values during the winter. This is quite usual in places with frost and snow on the ground. The frost makes it difficult for the gas to diffuse out of the surrounding ground and to be released directly into the air. Consequently, the radon gas is more likely to seep into the house. It is the radon concentration averaged over a long-time that is important with regard to health risk and, consequently, remediation. Decisions on what action to take are generally based on an average annual level. The type of equipment for radon measurements must have a long collection time. One method is to use *trackfilm-detectors*. The method is based on the fact that α-particles make tracks in certain materials such as polymers, minerals or glass. The tracks can be made visible with a microscope by etching. The detectors are usually very light (20 grams) and are well suited for large measuring programs. The collection time is usually from one month to about a year.

Another method uses thermoluminescence dosimeters (TLDs). This method consists of a small box with carbon powder and two TLD crystals. One crystal is placed in the middle of the box whereas the other is outside the box. Radon and radon daughters from the air will become trapped in the carbon powder and subsequently expose the crystal in the box. The crystal outside the box serves as a control for the cosmic radiation and gamma radiation from the walls of the room. The difference between the two TLD-crystals yields information on the average radon concentration during the exposure period.

There are also more direct instruments that make it possible to follow the radon concentration as a function of time, the concentration of the different radon daughters, as well as the fraction not bound to particles.

Radon doses

It is possible to make reasonably accurate measurements of the radon concentration (given in Bq/m^3) in a dwelling but, it is far more difficult to make estimates of the annual dose (given in mSv) to those living in a dwelling. Consequently, when an annual dose of 2 mSv is given, the value is based on an elaborate "dose model". This model considers how the radon daughters enter and exit the respiratory system. It is the short-lived decay products that yield the largest dose contribution. They are composed of elemental metals and can become fixed to the bronchia. These α-particle emitters cause damage to primitive lung cells. The dose received by these cells depends on whether the daughters are bound to dust particles or are free, with the free radon daughters yielding the largest dose.

When the radon atom emits an α-particle, Po-218 is in a free form. The radon daughters then bind to dust particles or to surfaces and furniture, removing them from the air.

An estimation of the radon dose from a given radon concentration takes into account both physical and biological parameters such as breathing frequency, way of breathing (nose or mouth), age, and sex. The calculated annual dose also takes into account the time spent indoors. The outdoor radon dose is much smaller.

In the thorium series there is another radon isotope (Rn-220) that is called thoron. This isotope also has decay products. Thoron decay products yield a dose that on average is approximately 10% of that from radon.

One key point made in this section is that, in large areas of the world, the two isotopes of radon and thoron together with their daughters yield an annual dose to the public of about 2 mSv (in the range 1 to 3). A second point is that, for areas where the dose level is large, there are simple and effective methods to reduce them. And third, it is important to remember that, in order to assess the health risk, the equivalent dose must be calculated from the absorbed dose using the radiation weighting factor. At the present time, the ICRP uses a w_R of 20 for α-particles.

Summary

In this chapter we have presented the four main natural radiation sources that give all of us a radiation dose from the time of conception until we die. The combined annual dose is around 2–3 mSv throughout the world, with a variation range of at least a factor of 10.

The most variable source is radon, varying between widely different extremes. Another important aspect of the radiation emitted by radon is that a large fraction consists of α-particles. Because α-particles produce a high density of ionizations along the track (high LET radiation, see page 32) and because the emitter is localized in the lungs, the effective dose from this source (given in Sv) is much larger than the actual dose (given in Gy).

Members of the public often seem to ignore natural doses of radiation but they are very concerned about smaller doses from anthropogenic sources such as nuclear accidents (Chernobyl, Three Mile Island, etc.) and radioactive waste products from the nuclear industry. In order to carry out epidemiological studies on small doses, the natural background doses are very important. The natural background not only produces about half a billion ionizations in your body per second, it does so continuously from "cradle to grave". This will be discussed in more detail in Chapter 11.

An important anthropogenic source is the use of radiation for medical purposes, industry and research. In the next chapter, we see that the medical use in all industrial countries contributes an annual dose of 0.4 mSv to greater than 1 mSv.

Chapter 8

Radiation in Medicine and Research

Artificial or anthropogenic (human made) radiation sources are used extensively in medicine, research and industry, and these sources are under regulatory control.

In this chapter and the next, the use of artificial radiation sources will be examined with a focus on estimating radiation doses to those working with these sources as well as others who may be exposed.

Radiation Sources

X-ray machines are by far the most numerous and significant of the artificial radiation sources. Hospitals throughout the world use different x-ray machines for many diagnostic purposes. X-rays are also important in the practice of dentistry and chiropractory. In addition, a number of hospitals have radiation producing equipment, such as linear accelerators, used for treating cancer.

It is important to note that the radiation from x-ray sources can be turned on and off. There are, therefore, no problems with storage and during periods when the equipment is not in use. When these devices are in use, the radiation field can be limited by lead screens and collimators.

X-ray diagnostics

A few months after the discovery of x-rays, the first x-ray pictures were published, showing the possibility of *seeing inside* a living human. On the left is shown one of the first X-ray pictures, taken in May 1896. On the right is a mammogram taken almost 100 years later.

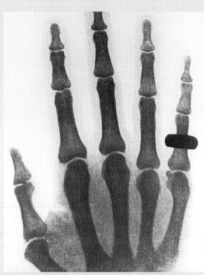

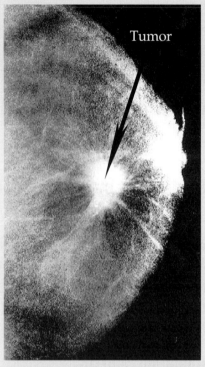

Tumor

Note the differences in these two pictures. In the picture of the hand, details of the bone structure and a ring are readily recognized. It is a lot more difficult in the mammogram to distinguish between cancer tissue and normal tissue. With knowledge about the absorption of x-rays, the equipment can be used to achieve this goal.

X-rays are absorbed more efficiently by heavy atoms than light atoms due to the increase in electron density (see Chapter 2). The large differences in electron density between bone and soft tissue are easy to see. The small difference in electron density between normal tissue and tumor tissue is more difficult to observe.

X-rays with low energies (about 30 kV) are used in mammography. The rationale for this is that soft X-rays are mainly absorbed by the photoelectric effect (see page 16), which is more sensitive to small variations in the electron density. Higher energy x-rays from a machine with a voltage of 100 kV are absorbed by the Compton process, which is not as sensitive to small changes in the electron density. As the energy increases more tissue can be penetrated and, for a picture of lung, a voltage of 100 kV is necessary.

Radioactive isotopes are used in medicine, research and industry. Some isotopes may be used for diagnostic purposes, whereas others are used for therapy. Some of the isotopes used for diagnostic purposes are: Tc-99m, I-131, Xe-133, Tl-201 and Au-195m. The isotopes are produced, transported to the institution involved and used by the clinician. Any radioactive wastes must then be stored in a safe way until the activities have decreased to an acceptable level.

X-rays in Medicine

X-rays are produced when high speed electrons are suddenly stopped (this radiation is sometimes called "bremsstrahlung"). In an ordinary x-ray tube, the electrons are accelerated by the voltage difference between the two electrodes in the tube (see illustration below). The voltage difference may vary between 20,000 volts and 300,000 volts (20–300 kV). The electrons then strike the anode, which consists of a heavy metal such as tungsten. After striking the anode, most of the energy of the electrons is given off as heat (the anode is usually cooled by circulating water) but a fraction is converted to x-rays. The maximum energy of the radiation x-ray photons is equal to the voltage between the electrodes.

If the voltage between the electrodes is in the range of 25 kV to 50 kV, they are called "soft" x-rays. Soft x-rays are used in mammography.

The x-ray picture. The principle for all diagnostic methods is that x-rays are capable of penetrating the body and interacting with electrons in the body (the interaction processes were described in Chapter 2). Regions with high densities of electrons absorb more of the x-rays than regions with low electron densities. It

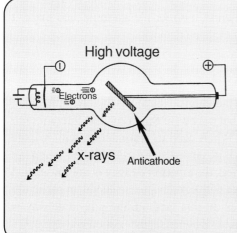

This drawing illustrates how x-rays are produced. The x-ray tube consists of an evacuated glass tube with two electrodes, the cathode and the anode. The voltage between the electrodes determines the type of x-rays produced. Electrons are released from the cathode, accelerated in the voltage gap and then strike the anode at high velocities The anode, frequently called the anti-cathode, consists of a heavy metal, such as tungsten. Part of the electron energy is dissipated as x-rays.

is the radiation that passes through the body that is observed on a film or fluorescent screen.

Therapy

In therapy, the purpose is to destroy cancer cells with radiation while sparing nearby healthy cells. This requires a careful balance between the benefit and the risk. Since the cancer cells are located inside the body, the radiation must pass through some healthy tissue before hitting the target. It is, therefore, a challenge to pick out the most suitable type of radiation and then decide upon an irradiation protocol. As you can see from Figure 6.5, the treatment requires high energy radiation that yields a suitable depth dose curve. Consequently, the therapy machines generate radiation with energies of 10 MeV to 30 MeV. The voltage between the electrodes in an ordinary x-ray tube can be hardly more than 300 kV because of electrical breakdown. When breakdown occurs, charges move between the anode and cathode in an uncontrolled manner, analogus to lightning striking. However, there are a number of other types of machines that are used for accelerating electrons up to high energies, such as the betatron and the linear accelerator. The use of linear accelerators for cancer treatment is now quite common.

In addition to the radiation sources used in medicine, there are a number of research accelerators as well as nuclear reactors. A few reactors are used for the production of radioactive isotopes which are used in medicine, research and industry.

In recent years, there have been large improvements in x-ray diagnoses due to the use of contrast agents and computer tomography (CT). Contrast agents are compounds that seek out the site of interest, a tumor for example, and make it more visible by virtue of having a high electron density.

Other Diagnostic Methods

Before leaving the discussion of medical radiological diagnosis, we briefly mention two types of *non-ionizing* radiation that penetrate the body and interact with tissue. One example is radio waves. When used in conjunction with a large magnet, the interactions of radio waves is observed by a method called *magnetic resonance imaging* (MRI). In this method, the electron density is not the critical variable because the radio waves interact with certain atomic nuclei, in particular the hydrogen atoms in water molecules. In this method, the proton density is observed. Furthermore, information can be obtained on the motion and dynamics of the water molecules.

Metastable isotopes used for medical diagnostics

Disintegration by a radioactive isotope starts with either an α- or a β-particle emmision. If the nucleus is still unstable, it emits γ-radiation immediately (in a fraction of a second). If this emission is delayed (for minutes or hours), it is a *"metastable"* isotope and this metastable property can be used for medical diagnostics.

The use of metastable isotopes

One metastable isotope is formed when molybdenum (Mo-99) emits a β-particle and is transformed to technetium (Tc-99m). It is customary to add "m" to the designation in order to point out that Tc-99m is a metastable isotope. Eventually it will emit γ-radiation, but because of the special structure of the nucleus, this emission is delayed by several hours (half-life of 6 hours).

This isotope is used in diagnoses in the following way:

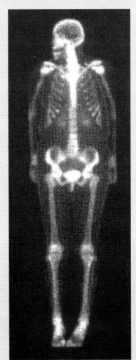

The starting point is Mo-99 bound to aluminum-oxide. When the compound is rinsed with physiological saline, any Tc-99m that has formed follows the water. Compounds that bind technetium are then added to the Tc-99m solution. The compounds are chosen according to their specificity for targets of interest. Common targets include the lungs, kidneys, or bone.

Tc-99m emits γ-radiation with an energy of 0.14 MeV, which readily escapes the body and is easily measurable. The distribution of radioactivity in the body can be measured with an instrument called a gamma camera. By comparing the picture obtained for a patient with that of a healthy person, information is obtained about the illness.

The method has several advantages compared to x-rays. The doses to both the patient and the medical personnel are small. The strength of the source used for an examination is around a few hundred million Bq. In the example to the right, 700 million Bq was used.

In this particular example, Tc-99m was added to methylene-diphosphonate, which is absorbed by the bone-forming cells (the osteoblasts). This kind of picture, called a whole body scan, makes it possible to study diseases of the skeleton, such as bone cancer.

Courtesy of Arne Skretting,
Norwegian Radium Hospital

Since MRI does not involve ionizing radition, its use lessens the average public dose by reducing the use of diagnostic x-rays.

A second example is ultrasonic waves. High frequency sound waves (which are quite different from electromagnetic waves) penetrate the body, bounce back, and are gathered to form an image. This method is commonly used for heart and pre-natal examinations.

Radiation Therapy

Shortly after the discoveries of Roentgen and Becquerel, it was evident that ionizing radiation could cause biological effects such as skin reddening, sore eyes, and loss of hair. Both Pierre Curie and Becquerel developed sores on their fingertips as a consequence of their work with radioactive materials.

H. Becquerel said in his Nobel lecture in 1903 that radium probably could be used to treat cancer. This turned out to be true and a number of hospitals started using radium for radiation therapy. Today radium is no longer used because of problems related to the radon gas that is formed. One thing retained from that period is the existence of treatment centers having the word radium in their names (for example Radiumhemmet in Stockholm, Sweden and Radiumhospita-let in Oslo, Norway).

Radiation therapy is one of the most powerful methods available for treatment of cancer, benefitting about 50% of all cancer patients. It is used, in combination with surgery and chemotherapy, as a primary mode of treatment and it is also used for palliative purposes. In a number of countries radiation is used extensively; unfortunately, there are still many countries where the use of radiation is far from ideal due to the lack of equipment and educated trained personnel.

As mentioned above, the type of radiation used is mainly x-rays from large therapy machines (mainly linear accelerators). In some cases, γ-rays from radioactive isotopes such as Co-60 and Cs-137 are used.

Research

Biophysics and biochemistry research laboratories use radioactive isotopes extensively. Researchers have learned a great deal about life processes by using radioactive isotopes bound to proteins, nucleic acids and their building blocks. By measuring the emitted radiations, researchers can follow isotopes and their reactions. This is called a "tracer technique", the compound is labeled and the fate of the compound is traced through its emission.

Table 8.1. Some isotopes and the *average* energy per disintegration

for their emitted β- and γ-rays

Isotope	Symbol	$t_{1/2}$	Energy γ in MeV	Energy β in keV
Tritium	H-3	12.35 years	–	5.68
Carbon-11	C-11	20.38 min	1.02	385.5
Sodium-24	Na-24	15.0 hours	–	553
Phosphorous-32	P-32	14.29 days	–	695
Sulfur-35	S-35	87.44 days	–	48.8
Strontium-89	Sr-89	50.5.days	–	48.8
Strontium-90	Sr-90	29.12.years	–	196
Molybdenum-99	Mo-99	66 hours	0.15	391
Ruthenium-103	Ru-103	39.28 days	0.468	74.5
Ruthenium-106	Ru-106	368.2 days	–	10
Iodine-123	I-123	13.2 hours	0.171	28
Iodine-131	I-131	8.04 days	0.38	190
Cesium-134	Cs-134	2.06 years	1.55	163
Cesium-137	Cs-137	30 years	0.626	187
Xenon-133	Xe-133	5.3 days	0.393	48.8
Barium-140	Ba-140	12.74 days	0.182	311
Cerium-141	Ce-141	32.5 days	0.0761	170
Cerium-144	Ce-144	284 days	0.0207	91

Some important isotopes in tracer techniques are given in Table 8.1. Note that the energies given in Table 8.1 represent *the average energy per disintegration*. In order to explain this in more detail, consider an example. The decay scheme for Cs-137 is given in Figure 2.6 showing that 94.6% of the disintegrations yield a γ-photon with an energy of 0.662 MeV. The average γ-energy per disintegration is consequently:

0.662 MeV · 0.946 = 0.626 MeV.

The average energy of the β-particles is approximately 1/3 of its maximum energy. Most references specify just the maximum energy.

Radioactive tracer techniques have given researchers opportunities to study the formation and breakdown of important biomolecules and to study the mechanisms underlying these processes. A long series of examples in which the tracer technique plays an important role could be given but instead we restrict ourselves to only one, the famous experiment of Alfred Hershey and Martha Chase (see next page).

The Hershey–Chase experiment

A famous experiment demonstrating the use of radioactive isotopes was carried out by Alfred Hershey and Martha Chase in 1952. They studied the mechanism for virus attack on a bacterial cell.

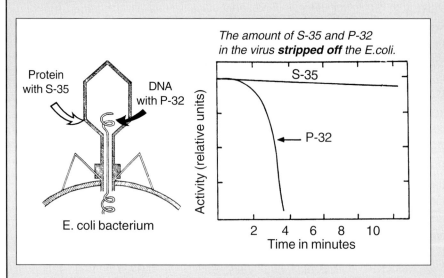

A virus consists of a cloak of protein which envelopes a nucleic acid (RNA or DNA). In this particular experiment, Hershey and Chase used a virus called T2 and an *E. coli* bacterium. T2 is a bacteriophage, a virus that infects bacteria. The protein making up the outer coat of T2 was labeled with the isotope S-35 and its DNA was labeled with P-32. Both are β-emitters but they have different energies and half-lives.

When the virus attacks the cell it becomes attached to the surface and after a few minutes the cell is infected. The question is: what is the mechanism for this process? Hershey and Chase worked out a technique that made it possible to "strip off" the virus from the cell. They used this technique and measured both the S-35 and P-32 activity in the virus that first became attached to the bacterium and then stripped off.

The figure demonstrates that the S-35 (or the protein) activity is almost constant, whereas the P-32 activity is rapidly lost after a couple of minutes. The DNA disappears from the virus that was subsequently stripped off. The explanation is that the DNA-part of the virus is injected into the cell and takes command of the bacterium. The protein envelope stays on the outside of the bacterium and that is stripped off.

This important experiment not only showed the time lapse of a virus infection but also that DNA contains genetic information; i.e., DNA is the molecule involved in heredity (see also Chapter 12).

Target-directed isotopes for radiation therapy

In radiation therapy, the purpose is to destroy cancer cells while protecting healthy cells as much as possible. In order to achieve this goal, one possibility is to bring the radiation source directly to the target (the cancer cells). This would increase the probability of hitting only the cancer target. The method presented here uses radioactive isotopes that are brought to the target with the help of antibodies. The possibility of hitting only cancer cells is improved if the source emits α- or β-particles, since these particles deposit energy to a very small region.

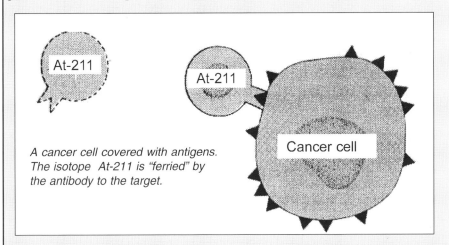

A cancer cell covered with antigens. The isotope At-211 is "ferried" by the antibody to the target.

In order to irradiate the thyroid, radioactive iodine (I-131) is often employed. The body itself will transport the isotope specifically to the thyroid, which is then irradiated by short range β-particles. This means that only thyroid cells (and cells nearby) are damaged, acheiving the goal of the procedure.

Isotopes emitting α-particles may even be better suited for the purpose. One is an astatine isotope, At-211, which has a half-life of 7.2 hours. The idea is to use this isotope and employ a transport-system that brings the isotope close to the target cells. *How can this be done?*

Antibodies can be used as "transporters"! One of the requirements for this method is that the cancer cell in question have a specific antigen on its membrane surface. The antibody to this antigen must be produced and the radioactive isotope attached. The drawing above illustrates the method. The antibody brings the isotope At-211 to the cancer cell and binds to the antigen. A disintegration, which includes an α-particle, has a considerable chance of damaging only the target cell.

This particular "transport system" can also be used for other medicines or fluorescent compounds.

Isotopes Used in Industry

Radiation sources can be used for a number of purposes in industry, such as in *industrial radiography*. The method is based on the same idea used in medical diagnosis. The aim is to "see" into the interior of a material; for example, to examine welding connections and/or cracks in a structure. For this purpose, γ-rays from radioactive isotopes (often Ir-192) and x-rays are used.

The radiation sources used in industry usually have very high activities. Ir-192 sources, on the order of 1.5 TBq (one million million Bq $=10^{12}$ Bq = 1 tera becquerel = 1 TBq), are used. Even larger sources may be used for some purposes. For example, the "Liberty Bell" in Philadelphia was studied using a Co-60 source of several hundred TBq to discover faults that could not be seen otherwise. Another example is the use of a 1 MeV x-ray machine (in the 1940s) to produce an x-ray film of an entire jeep.

A different use of radioactive sources is for process control. One simple example is to control the level in a storage tank, for example grain in a silo. A γ-ray source is mounted on one side of the silo and a detector on the other side. As long as a signal is detected, there is air between the source and the detector. When the signal decreases, the grain has reached the level of the detector and reduces the number of γ-rays hitting the detector. The sources used are Cs-137 or Co-60. By connecting the detector to a mechanism one can stop the filling of the silo when a predetermined level has been reached. Optical instruments in the same situation are ineffective because they become covered with dust.

When radioactive tracers are used in industry, an effort is made to use isotopes with short half-lives in order to minimize the waste problem.

Smoke detectors in our homes utilize radioactivity. They consist of a radioactive source in an ion chamber. Since the radiation ionizes the air in the ion chamber, a small electric current is produced. When smoke particles enter the chamber, the electric current is drastically reduced and the alarm turns on. Because the detectors use α-emitters (usually 40 kBq of Americium-241), no radiation can be detected outside the chamber.

If a radioactive compound is mixed with a fluorescent compound, a self-luminous compound is formed. This was used in exit signs in industry. It was previously noted that, for this purpose, radium was used and painted on numbers and pointers on clocks and instrument panels. Due to the penetrating nature of γ-radiation, radium is no longer used; isotopes that only emit β-particles have been substituted. The β-particles have such a short range they do not make it into the air.

The old radium watches contained activities of about 5,000 to 10,000 Bq. Measurements indicate that such watches gave an annual dose to the skin (under the watch) of 1.3 mGy and to the gonads, about 0.1 mGy per year. The main health risk presented by these watches was that the radium released small amounts of radon.

Sterilizing Medical Equipment

Large radiation doses will kill bacteria, fungi and insects. For such purposes, Co-60 sources (1,000 to 10,000 TBq) emitting γ-rays (with energies 1.17 and 1.31 Mev) are used.

Radiation has been used since 1958 for sterilizing medical equipment that other-wise was difficult to sterilize by heat and steam, such as syringes, bandages, blood transfer equipment and a large number of other health care instruments and materials. The radiation doses used must be sufficient to inactivate bacteria and viruses and are of the order 20 to 40 kGy. There is no doubt that this method of sterlization has played an important role in reducing the incidence of infections. These are important examples of the benefits derived from the use of radiation.

Irradiation of Food Products

A great leap of the imagination is not required to go from using radition to sterilize medical equipment to sterilizing food. Irradiation of food began in the early 1950s. The method is useful even in cases where the food products can be preserved by freezing or with chemical additives. For many food products, such as fish, poultry, vegetables, fruits and spices, irradiation increases safety and does a superior job of preserving quality. The doses used for food products (5 to 10 kGy) are generally smaller than those used for sterilizing medical equipment.

Irradiation of food has been shown to be both safe and effective. It is probably the best method devised for food preservation since the tin can (made of tinned iron or other metal) was introduced in 1810. Irradiation is now one of the preferred tools for eliminating infectious microbes, such as Salmonella. With the increased use of radiation, the use of chemical preservatives, such as ethylene oxide and ethylene bromide, can be reduced. This is advantagous because irradiation leaves food much closer to its natural state than do chemical preservatives.

In some European countries you can buy milk that will stay fresh for months. The milk itself is treated with UV-radiation, whereas the plastic containers are γ-irradiated. Milk is not suited for γ-irradiation because the taste changes at rather low doses. For most other foods, change in taste is not a problem. Indeed, the taste and freshness of strawberries are superior when radiation is used for preservation.

Some people are afraid of negative effects of food irradiation and some environmental organizations claim that a health risk may exist. Proteins and nucleic acids are damaged by radiation. Chemical bonds are disrupted, free radicals are formed, and ultimately new chemical derivatives are produced. In order to answer such concerns, the following results, stemming from extensive research, should be considered:

- Irradiation of all foods at properly controlled doses results in very minor chemical changes. In almost all foods (shell fish are a notable exception), the chemical changes are not harmful to humans.
- Irradiation does not result in changes of nutritional quality. The influence is smaller than for a comparable heat treatment.
- Since external γ-radiation from Co-60 or Cs-137 is used, **no** radioactivity can be induced in the food.

The World Health Organization (WHO) and a number of national health organizations have permitted the use of irradiated food products. The method is safe, very effective, economical, and readily monitored. Although the dose used could be presented on an information label, it would perhaps be more informative to give some measure of new chemicals produced. The latter would make clear that the amount of unnatural chemicals produced by radiation preservation is minute compared to the use of chemical perservatives.

Control of Insects

Large quantities of chemicals are used in agriculture to control insect infestations. The chemicals are usually poisonous and represent a biological risk. An alternative to these chemicals, in some cases, is the use of radiation. The following procedure can be used.

First, a certain number of the particular insect of interest are collected. Then one breeds the insects in order to increase their population. Before these insects are released, they are γ-irradiated in order to make them sterile. When the sterile irradiated insects mix with unirradiated insects, they are no longer able to breed. There are problems connected with this method; the number of irradiated insects must be sufficient and of the same order of magnitude as the wild insects and the method is specific since it is valid only for the type of insect collected.

Doses Due to the Medical Use of Radiation

Annual dose = 0.6 mSv

The dose due to medical uses of radiation adds to the ever-present background dose due to natural sources. The individual dose due to medical applications may vary from zero (some people have never been examined by x-rays) to hundreds of mGy.

There have been two trends in recent years, working in opposite directions. Increases in the average annual dose are due to the introduction of new applications and due to the increased use of existing applications. Decreases in the average dose are due to new equipment and improved methodology. An example of these two trends can be seen in mammography. The number of patients and frequency of exams has increased while the dose required for each exam has been reduced. It is now recommended that all women above a certain age have regular mammography examinations. While the exam frequency depends on such factors as family history, most women above 50 years of age get a mammogram once per year.

It is possible to determine the dose for each type of examination but, it appears that, the same examination in two different hospitals may differ considerably. It is, therefore, difficult to estimate the average annual dose due to medical use of radiation.

The United Nations Subcommittee on the Effects of Atomic Radiation (UNSCEAR) has made some effort to estimate an average annual dose to the public due to medical uses of radiation. They estimate that for industrialized countries the dose is approximately 1 mSv per year. For the total world population the average annual dose is estimated to be 0.4 mSv. In some countries, France for example, the average annual dose is a bit larger due to a higher frequency of liver and kidney exams. For Scandinavia the average annual dose is considered to be 0.6 mSv.

Nuclear weapon tests

The first nuclear bomb test took place near Almagordo in New Mexico in July of 1945. Since then, the United States, the Soviet Union, England, France, China and India have tested more than 1,900 weapons in the air, on the ground and underground. The map below shows most of the places used for these nuclear tests.

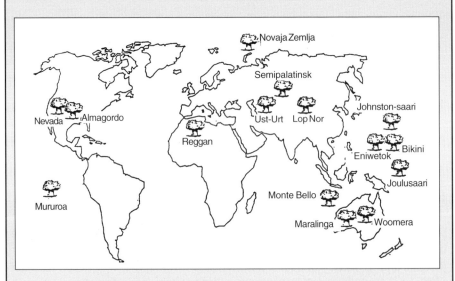

The release of radioactive isotopes depends on the type of bomb and, most of all, whether the bomb is detonated in the air, on the ground or underground. There is a considerable difference between a blast in the atmosphere and an underground explosion. Only the atmospheric tests result in the fallout of radioactive isotopes.

Furthermore, a distinction should be made between explosions near the ground and explosions that take place at altitudes where the so-called "fire ball" does not reach the ground. The largest amounts of radioactive isotopes in fallout are found when the explosions take place near the surface.

Most atmospheric nuclear tests took place in the years up to 1963. France and China performed atmospheric tests up to 1978 and 1980 respectively.

The underground tests on Novaja Zemlja started in 1964, and up to now approximately 40 tests have taken place. France performed tests at Mururoa in 1995.

The radiation doses to the public from all these nuclear tests have for the most part been very small. They cannot be measured against the natural background doses. The exceptions are a few atmospheric tests performed in the early years.

Chapter 9

Nuclear Weapons and Reactor Accidents

Nuclear Bomb Tests

This chapter is concerned with radiation doses to the public from nuclear weapons tests, as well as those resulting from nuclear reactor accidents that have occurred over the years. Since the doses involved are mostly small (smaller than the doses from natural radiation), it is extremely difficult to pinpoint the health effects from these extra doses. This is a widely debated issue and will be discussed in more detail in Chapters 11 and 12. Here we will concentrate on the doses.

During the period from 1945 to 1981, 461 nuclear bomb tests were performed in the atmosphere. The total energy in these tests has been calculated to be the equivalent of about 550 megatons of TNT (TNT is the abbreviation for trinitrotoluene). The bombs in Hiroshima and Nagasaki had a blasting power of, respectively, 15 and 22 thousand tons of TNT. Nuclear tests were particularly frequent in the two periods from 1954 to 1958 and 1961 to1962.

Several nuclear tests were performed in the lower atmosphere. When a blast takes place in the atmosphere near the ground, large amounts of activation products are formed from surface materials drawn up into the blast. The fallout is particularly significant in the neighborhood of the test site. One of the best known tests with significant fallout took place at the Bikini atoll in the Pacific in 1954 (see next page).

A bomb test in the Pacific

On March 1, 1954, the United States detonated a hydrogen bomb (with a power of about 15 million tons of TNT) at the Bikini-atoll in the Pacific. The bomb was placed in a boat in relatively shallow water. Considerable amounts of material (such as coral) were sucked up into the fireball and large amounts of activation products were formed.

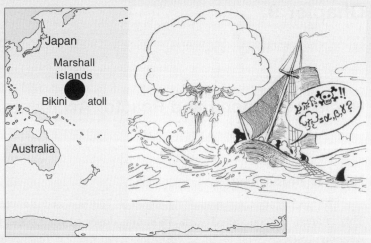

A couple of hours after the blast, the instruments on the American weather station on Rongerik island (about 250 km away) indicated a high radiation level. The radiation increased rapidly and it was decided to evacuate about 280 people living on the neighboring islands; Rongelap, Alingiae and Utirik. Because the fallout for these islands was so large, the inhabitants were not allowed to live there for 3 years.

Approximately 130 km from the test-site was the Japanese fishing boat Fukuru Maru with 23 fishermen aboard. After the blast they pulled in the fishing equipment and sailed away. Approximately four hours later, the fallout started in the area where the boat had moved.

Dust, soot and even larger particles came down. The crew lived with this for a number of days and took no special precautions with regard to hygiene, food, and clothing since they had practically no knowledge of radioactivity and its biological effects.

The fishermen received very large doses, about 2 to 6 Sv. They felt nauseous and received skin burns from β-particles in the fallout. One of the fishermen died within 6 months, but radiation was probably not the cause of death. Most of the fishermen were still alive 30 years later. Chromosome analyses showed larger amounts of damage than normal in their lymphocytes. The importance of the damaged lymphocytes is covered in Chapter 12.

Because of the extreme temperature of a nuclear explosion, the radioactive material becomes finely distributed in the atmosphere. A certain fraction is kept in the troposphere (the lower 10 km) and is carried by the wind systems almost at the same latitude as the explosion. This part of the radioactive release will gradually fall out, the average time in the atmosphere being about one month.

The main fraction of the radioactive debris from an atmospheric test goes up into the stratosphere (10 to 50 km).This can remain in the stratosphere for years since there is a very slow exchange between the troposphere and the stratosphere. The fallout consists of several hundred radioactive isotopes; however, only a few give significant doses. The most important are listed below.

- **Zirconium-95** (Zr-95) has a half-life of 64 days and **iodine-131** (I-131) has a half-life of 8 days. Both of these isotopes, in particular I-131, are of concern for a short period (a few weeks) after being released to the atmosphere.

- **Cesium-137** (Cs-137) has a half-life of 30 years. The decay scheme for this isotope (Figure 2.4) shows that both β-particles and γ-rays are emitted. The β-emmision has an impact on health when the isotope is in the body or on the skin. The γ-radiation has an impact both as an internal and external radiation source.

- **Strontium-90** (Sr-90) has a half-life of 29.12 years. This isotope emits only a β-particle and is difficult to observe (maximum energy of 0.54 MeV). This isotope is a bone seeker and is important when the isotope enters the body. It should be noted that Sr-90 has a radioactive decay product, Y-90, which has a half-life of 64 hours and emits β-particles with a maximum energy of 2.27 MeV. With this short half-life, it is likely that this amount of β-energy will be deposited in the same location as those from Sr-90.

- **Carbon-14** (C-14), while not a direct product of fission, is formed in the atmosphere as an indirect product. The fission process releases neutrons that interact with nitrogen in the atmosphere and, under the right conditions, C-14 is formed as an activation product. The individual doses from this isotope are extremely small. However, due to the long half-life of 5,730 years, it will persist for many years. When C-14 is used in archeological dating, it is necessary to correct for the contribution from the nuclear tests.

Nuclear tests at Novaja Zemlja in 1961 and 1962

In 1961 and 1962, a number of atmospheric nuclear tests took place at Novaja Zemlja. The tests have been of great concern for people living in the northern hemisphere, in particular, Scandinavia. The fallout, which was largely determined by precipitation, was quite large on the western part of Norway as illustrated below. The isotopes Cs-137 and Sr-90 then entered the food-chain via grass (in particular reindeer lichen). Consequently, sheep, cows and reindeer ingested radioactive material when feeding on grass and reindeer lichen. People eating the meat or drinking the milk from these animals received some extra radioactivity.

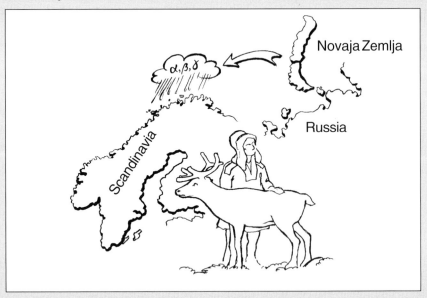

Many measurements were carried out in order to determine the activity and types of isotopes in the food products.

Mainly, scintillation counters were used and the observations were concentrated on the γ-radiation from Cs-137. It is far more difficult to observe Sr-90 since it only emits β-particles. Attempts were made in particular experiments to measure the ratio between Cs-137 and Sr-90. This ratio was assumed to be rather constant implying that the Cs-137 observations also yielded information on Sr-90.

The Cs-137 activity in food products (meat, milk, cheese, etc.) was measured. Furthermore, whole-body measurements were started. The latter were performed using large scintillation crystals placed above the stomach. It appeared that Cs-137 entered the body and can be found in all of us. A few examples are given in Figures 9.2 and 9.3.

Nuclear Tests on Novaja Zemlja

The nuclear tests of most concern for the Northern Hemisphere were performed by the former USSR (Russia) on the island Novaja Zemlja located in the Arctic, approximately 1,000 km from northern Norway. When these islands were chosen as a test site in 1954, more than 100 families lived there. They were all removed from their homes. Altogether 87 atmospheric nuclear tests were performed at this site. The activity was particularly large during 1961 and in the fall of 1962. Most of the tests were performed at high altitudes, thus the "fireball" did not reach the ground. Consequently, the production of activation products was limited.

However, the radioactive debris from the tests was released into the atmosphere. Calculations indicate that the atmospheric nuclear tests (including those from United States, England, France and China) have yielded a total release of Cs-137 of 1.0–1.4 million TBq (a TBq is 10^{12} Bq), or approximately 30 million Ci. The total release of Cs-137 from all the bomb tests is approximately 30 times larger than that released during the Chernobyl accident. The total release of Sr-90 is calculated to be about 0.6 million TBq (approximately 75 times larger than the Chernobyl accident).

As mentioned above, when a blast takes place in the atmosphere, a large fraction of the radioactivity will go through the troposphere and into the stratosphere. Since the exchange between the two is rather slow the radioactivity will remain in the stratosphere for a long time. Westerly winds will bring the activity to the east. The radioactivity from the nuclear tests in the 1960s was distributed over large areas; however, the amount of fallout varied from one region to another according to the variation in rainfall (most of the fallout came down with the rain). The fallout pattern from the nuclear tests was different from that of the Chernobyl accident, which was much more dependent on the wind directions since the release itself was restricted to the troposphere.

From September 10 to November 4, 1961, the Soviets carried out 20 nuclear tests at Novaja Zemlja. The power of the bombs varied from a few kilotons TNT (equal in power to Hiroshima bomb) to approximately 58 megatons TNT, which is probably the largest bomb ever detonated. The release of fission products to the atmosphere was large and could be observed for long distances from the test site. For example, in Oslo, Norway (about 2,000 km away), an increased level of radioactivity in the air was observed (see Figure 9.1). These concentrations of radioactivity were measured simply by drawing air through a filter. Radioactive isotopes attached to dust particles in the air became absorbed on the filter (see picture on page 99). The radioactivity on the filter was measured, and since the air volume drawn through the filter was known, the activity could be calculated in Bq per cubic meter.

As can be seen in Figure 9.1, the activity started to increase on September 14 (4 days after the first blast). In October, the air activity 2000 km away was approximately 30 times larger than normal.

Similar measurements were performed in 1962. On November 7th, the air activity in Oslo was about 200 times above normal, indicating that one of the bombs (classified as middle power) which exploded on November 3 or 4, produced large quantities of fission products.

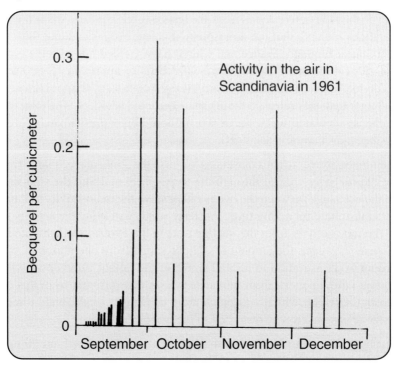

Courtesy of Anders Storruste, Inst. of Physics, Univ. of Oslo

Figure 9.1. The measurements presented here serve as an example of airborne radioactivity in combination with nuclear tests in the atmosphere. The data refer to the Russian nuclear tests on Novaja Zemlja in 1961. The measurements were carried out about 2,000 km away from the test site. The activity is given in Bq per cubic meter air.

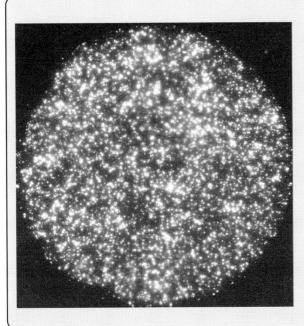

A radioactive filter

The radioactivity in the air during the nuclear tests at Novaja Zemlja in 1961 was measured by sucking air through a filter. The filter itself was laid directly on an x-ray film, and the white dots indicate small particles containing radioactive isotopes. The filter to the left is taken from an experiment carried out 2,000 km from the test site. The radioactivity reached the area after 4 days. The types of isotopes in the filter were measured with a scintillation counter.

Courtesy of Anders Storruste, Inst. of Physics, Univ. of Oslo

Radioactivity in Food

In the years since the bomb tests in the atmosphere were canceled, the amount of radioactive isotopes have continued to diminish. The fallout is dominated by the two isotopes Cs-137 and Sr-90. The fallout has decreased considerably since the mid-1960s but still, more than 30 years later, a small fallout persists from the bomb tests.

The radioactive isotopes hitting the ground become bound to plants, grass and, in particular, reindeer lichen. The activity in this plant decreases more slowly than that for plants withering in the fall.

The radioactive isotopes on the ground slowly diffuse into the soil. Some of them are taken up in plants via the roots. Consequently, a certain fraction of the fallout will find its way into the food chain and finally into humans. In addition to containing natural radioactive isotopes, many food products will also contain a small contribution from the fallout activity, mainly Cs-137. An interesting example of radioactivity in food is given in Figure 9.2.

This figure shows the activity of Cs-137 in reindeer meat. Many of the people living in that area eat reindeer meat every day and, consequently, they have a measurable content in their bodies. For a group of 20 people, the average activity was measured using whole-body counters over a period of more than 20 years. The results are given in Figure 9.2.

As can be seen, the activity has decreased slowly since the tests in the atmosphere ceased until the end of the period shown. After the Chernobyl accident in 1986 the activity increased due to new fallout.

Based on the results in Figure 9.2, it is possible to estimate the extra radiation doses as well as the ecological half-life for this area. The observations can be

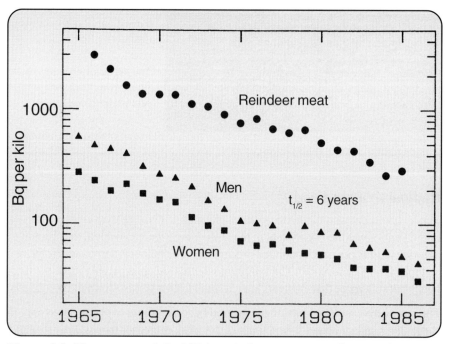

Figure 9.2. The content of Cs-137 in reindeer meat as well as in the people who own the animals. The example is taken from northern Norway. The activity is assumed to be evenly distributed in the body and is therefore given as Bq/kg. The reason for the difference between women and men is presumably the same as that for the content of K-40 (see Figure 7.3, page 71). Potassium and cesium are in the same column of the Periodic table and may be distributed in the body in the same way with a higher content when the muscle mass is large relative to the total mass. The ecological half-life (see page 24) is about 6 years. (Data courtesy of A. Westerlund, Norwegian Radiation Protection Authority)

fitted reasonably well to a straight line in the plot, implying that the activity decreases exponentially. The half-life is about 6 years for both the reindeer meat as well as for the people.

Looking at other groups of people with a different diet, the amount of activity due to the nuclear tests appears much smaller. In Figure 9.3 some data from Sweden, observed by whole body measurements, are presented (R. Falk, Swedish Radiation Protection Institute, SSI). A group of people from the Stockholm area have been followed since 1959. The measurements, therefore, include the effect of both the bomb tests of the 1960s and the Chernobyl accident in 1986. Furthermore, two groups (farmers and non-farmers respectively) from Gävle have been studied. Gävle is an area, north of Stockholm, which had the highest fallout (approximately 85 kBq/m^2) in Sweden from the Chernobyl accident.

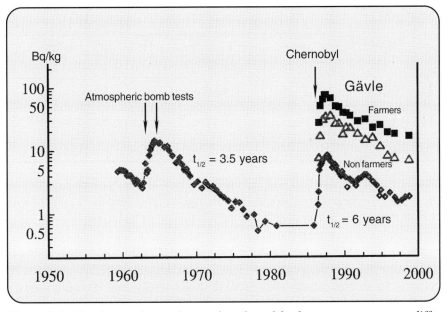

Figure 9.3. The figure shows the results of total body measurements on different groups of people in Sweden. (Data courtesy of R. Falk, Swedish Radiation Protection Institute, SSI)

As you can see, the total body activity for the Stockholm group reached a peak in 1965 (about 13 Bq/kg), which is a factor of 30–50 smaller than that of the Lapps (Figure 9.2). The data in Figure 9.3 can almost be fitted by straight lines and consequently half-lives can be caluculated. These half-lives may be considered as ecological half-lives and some values are given on the figures.

The data presented in the two figures also yield opportunities to make a rough calculation of the doses involved. Thus, we can estimate the dose obtained for the peak year (1965 for the bomb tests and 1986 for the Chernobyl accident), as well as the accumulated dose for the first 10 years (1965–1975 for the bomb tests and 1986–1996 for Chernobyl fallout). The data for the groups in Figures 9.2 and 9.3 are given in Table 9.1.

Table 9.1. Cs-137 doses due to the atmospheric bomb tests and the Chernobyl accident

Group	Bomb tests		Chernobyl	
	Dose peak year	Dose over 10 years	Dose peak year	Dose over 10 years
Lapplanders	1.5 mSv	8.8 mSv		
Gävle farmers			0.2 mSv	1.2 mSv
Stockholm group	0.03 mSv	0.14 mSv	0.03 mSv	0.18 mSv

The internal doses due to Cs-137 in the Lapps in northern Norway were among the highest to any group of people and very much higher than that to other members of the public. According to Figure 9.2, the Lapps had a whole-body activity in 1965 of approximately 600 Bq/kg for men and 300 for women corresponding to an equivalent dose of 1.5 mSv for men and 0.7 mSv for women that year. This extra dose in the peak year was approximately half that obtained by commercial air crews every year. From the bomb tests over a 10 year period the dose to the Lapplanders was approximately 8.8 mSv, whereas the dose to the Stockholm group was about 0.14 mSv. The dose from the natural background was about 30 mSv for the same period.

The dose figures for the Stockholm group would be equal to or larger than the dose to the average person on the Northern hemisphere (see page 78).

On the following pages we describe in more detail how it is possible to calculate radiation doses from radioactive isotopes in the body. The calculations are not exact but give a good overview of the doses involved. Those not interested can skip this section.

Radiation Doses from Cs-137 in the Body

The effects of nuclear bomb testing as well as the reactor accident in Chernobyl have been discussed. Now we shall describe in some detail how the doses can be estimated and then apply the calculation for the groups presented in figures 9.2 and 9.3.

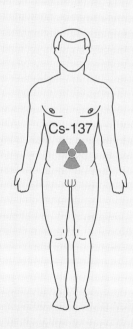

A radiation dose is, by definition, the energy deposited in the body. For radioactive isotopes we can estimate the energy deposited when we use the decay scheme. The decay scheme for Cs-137 is given in Figure 2.4. For every disintegration both a β-particle and γ-radiation are emitted. The energy given off into the body consists of the following:

β-particles

The β-particles have a very short range in tissue and will consequently be absorbed completely in the body. The average β-energy (E_β) is approximately 1/3 of the maximum energy given in the decay scheme. The following calculation is used (see also Figure 2.4):

$$E_\beta = 1/3 \ (94.6 \ \% \cdot 0.512 \ \text{MeV} + 5.4\% \cdot 1.174 \ \text{MeV})$$
$$= 0.183 \ \text{MeV}$$

This means that the β-particles from Cs-137 deposit about 0.18 MeV per disintegration.

γ-radiation

The γ-radiation will be partly absorbed in the body and partly escape from the body. It is the part of the γ-radiation that escapes from the body that is used in the measurements presented in Figures 9.2 and 9.3.

The γ-radiation from Cs-137 has an energy of 0.662 MeV. The radiation is absorbed according to an exponential function. A layer of about 8 cm of soft tissue will stop half of the radiation from Cs-137. *The half-value layer* in water and tissue for this γ-energy is approximately 8 cm. A rough estimate is, therefore, that approximately half of the γ-radiation from Cs-137 is deposited in the body (i.e. E_γ is about 0.33 MeV per disintegration).

The total energy deposited in the body per disintegration is the sum of the energies from both the β-particle and the γ-radiation, 0.18 MeV plus 0.33 MeV, giving

$$E = E_\beta + E_\gamma = 0.183 \text{ MeV} + 0.33 \text{ MeV} \approx 0.5 \text{ MeV}.$$

Dose

The radiation dose is the energy deposited per unit mass, measured in J/kg. Cs-137 is evenly distributed in the body, and the energy deposited per kg would be the number of disintegrations multiplied by 0.5 MeV. If we assume that the body burden is n Bq/kg and constant throughout a full year, the total number of disintegrations (N) would be n times the number of seconds in a year:

$$N = n \cdot 60 \cdot 60 \cdot 24 \cdot 365 Bq / kg \approx 3.15 \cdot 10^7 \cdot nBq / kg$$

The radiation dose is the product of the number of disintegrations and energy deposited per disintegration (remember that $1 \text{ eV} = 1.6 \cdot 10^{-19}$ J):

$$D = 3.15 \cdot 10^7 \cdot n \cdot 0.5 \cdot 10^6 \cdot 1.6 \cdot 10^{-19} J / kg = n \cdot 2.52 \cdot 10^{-6} Gy$$

Since the radiation consists of γ-radiation and β-particles with a radiation weighting factor of 1, the dose would be the same in Sv.

Returning to figures 9.2 and 9.3, we see that the Lapplanders in 1965 had a body burden of 600 Bq/kg. The dose that year was, therefore, 1.5 mSv for men and about half that value for women. The peak year doses for the other groups are given in Table 9.1.

Accumulated doses

As seen from the curves in Figures 9.2 and 9.3, the activities, and therefore the doses, decay exponentially. Since we roughly know the half-life, it is possible to estimate the total dose for 10 years. If the activity decays in a similar way, we can calculate accumulated doses accordingly. The accumulated dose for 10 years is found by the formula:

$$D_{total} = \int_0^{10} D_0 \cdot e^{-\lambda t} dt = (D_0 / \lambda)(1 - e^{-10})$$

Here D_0 is the first year dose, $\lambda = \ln 2 / t$ where t is the half-life in years. Using this formula, the doses presented in Table 9.1 are obtained. These are doses in addition to the doses from natural sources. The background radiation dose for a 10 year period in Scandinavia and most of the world is around 30 mSv.

The Chernobyl Accident

The exact amount of radioactive isotopes released during the Chernobyl accident is not known in detail. According to early reports, the release was approximately as given in Table 9.2.

Table 9.2. **The release of radioactive isotopes from the Chernobyl accident. The amount given in TBq(10^{12} Bq)**

Isotope	Half-life	Amount (TBq)
Cs-134	2.06 years	19,000
Cs-137	30.0 years	38,000
I-131	8.04 days	260,000
Xe-133	5.3 days	1,700,000
Mo-99	2.8 days	110,000
Zr-95	64 days	140,000
Ru-103	39 days	120,000
Ru-106	368 days	60,000
Ba-140	12.7 days	160,000
Ce-141	32.5 days	100,000
Ce-144	284 days	90,000
Sr-89	50.5 days	80,000
Sr-90	29.2 years	8,000

Summary Report on the Post-Accident, Safety Series No. 75,Vienna (1991)

Chernobyl is in the Ukraine, very near the border of Belarus. Approximately half of the released activity fell out in the area around the reactor. All of the plutonium and most of the strontium (Sr-89 and Sr-90) fallout was restricted to a region within 30 km of the reactor. However, for the cesium isotopes Cs-134 and Cs-137, the distribution was extensive. Belarus and the western parts of Russia received most of the cesium fallout, but considerable amounts were transported by the wind to western Europe.

During the first days after the accident, the wind direction was to the northwest (towards Scandinavia). Considerable amounts of fission products were transported to the middle regions of Sweden and Norway. Unfortunately, it was raining in some of these areas and the fallout was consequently large. Thus, in parts of Sweden (the area around Gävle, north of Stockholm) and in Norway the fallout of Cs-137 reached up to 100 kBq/m² (about 3 Ci/km²). The average value, however, was much smaller and on the order of 5 to 10 kBq/m².

During the first days, I-131 was the most significant isotope. Due to the short half-life of 8 days the activity decreased rapidly. As can be seen in Figure 6.4, this isotope was easily observed during the first phase after the accident.

Radioactivity in Food

Outside the former USSR, the radiation doses due to the Chernobyl accident can mainly be ascribed to the radioactivity in food products, in particular meat (from sheep and reindeer). The activity was dominated by the two cesium isotopes. Due to the short half-life of Cs-134 of 2 years, the activity decreased rapidly during the first years. Shortly after the accident, the activity ratio between Cs-134 and Cs-137 was approximately 1 : 2.

The average equivalent dose to people in Scandinavia was approximately 0.2 mSv the first year after the accident. About 2/3 of the dose was due to food products, and about 1/3 was due to external γ-radiation.

The radioactivity from Chernobyl will gradually decrease and the extra doses to the public will go down as the years pass. Figure 9.3 yields good information on the body burden of Cs-137 for different groups in Sweden. The estimates carried out indicate that the accumulated dose to people in Europe (for example Table 9.1) will be about 1 mSv in 50 years. In the same period the doses from natural background sources and medical use would be on the order of 200 mSv.

Release of radioactivity from Chernobyl

The Chernobyl nuclear reactor accident took place on April 26, 1986. The explosion and fire released a large amount of radioactive isotopes. The release was significant in the first 10 days, as shown in the figure below.

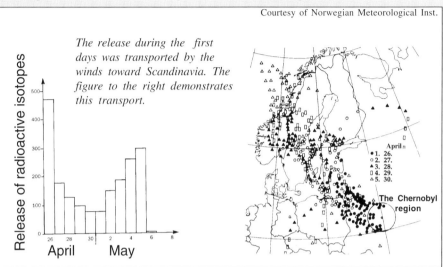

Courtesy of Norwegian Meteorological Inst.

The release during the first days was transported by the winds toward Scandinavia. The figure to the right demonstrates this transport.

Large amounts of radioactive isotopes were released into the atmosphere. The isotopes reached to an altitude of more than 2,000 meters. The compounds were then transported by the wind. The map demonstrates how the isotopes moved during the first days after the accident. The wind direction was toward northwestern Europe. Since it was raining at the same time in Scandinavia, the fallout became significant in some regions of Sweden and Norway. Other countries, such as Denmark, the Netherlands, Belgium, France and England received small amounts of activity.

The wind then turned South and countries like Romania, Bulgaria, Greece and Czechoslovakia had significant fallout during the latter part of the accident. The fallout of some isotopes, such as Sr-90 and plutonium, was mainly restricted to regions in the neighborhood of the reactor (within 30 km).

The fallout from the Chernobyl accident is now well known and maps are available which show the concentration (given as Bq/m^2 or sometimes as Ci/km^2) of the different long-lived isotopes such as Cs-137, Sr-90 and plutonium. It is, therefore, possible to estimate annual doses as well as lifetime doses for those living in the area (examples are given above). It is, however, more difficult to determine the doses from the short-lived isotopes such as I-131 as well as the acute doses to those at Chernobyl fighting the fire and cleaning the area.

Pollution Around Chernobyl

Most of the fallout was localized to the region in the neighborhood of the reactor. During the accident, radioactive isotopes were both in the air and on the ground. This resulted in a radiation level which made it necessary to evacuate about 130,000 people. The doses to these people are not known, but people in the countryside (within 15 km of the reactor) received the largest doses.

During the first year after the accident more than 200,000 *liquidators* worked on stopping the fire and cleaning the area. Some of these liquidators probably received significant radiation doses.

In 1989, the Soviet Union started to release information about the geographical distribution of radioactive isotopes. The information was not in agreement with a number of "reports" describing how radiation from the accident had resulted in poor public health. Some of these reports from the accident area were in disagreement with our knowledge on the biological effects of radiation. In 1989, the government in the former Soviet Union asked for international help to determine dose levels and health effects.

In the spring of 1990, a group of experts formed the *International Advisory Committee* with the purpose of studying the situation in the polluted regions of Ukraine, Belarus and Russia. The committee would evaluate the data released by radiation experts in the Soviet Union on pollution, doses and health effects.

The Committee consisted of a total of 200 experts from 23 nations with Itsuzo Shigematsu from Japan as chairman. He has, for a number of years, been head of the *Radiation Effects Research Foundation* in Hiroshima (they have worked with those irradiated in Hiroshima and Nagasaki). Several international organizations such as the World Health Organization (WHO), the International Atomic Energy Agency (IAEA), the UN committee for food and farming (FAO), the UN Scientific Committee on the Effects of Atomic Radiation (UNSCEAR), the International Labor Organization (ILO) and the World Meteorological Organization (WMO) were represented on the Committee. Laboratories in Austria, France and US took part in the analyses.

The organization made about 40 visits to the Soviet Union. This is one of the most ambitious international projects ever carried out in the radiation field. The report was released in May, 1991 (5 years after the accident). The conclusions can be summarized as follows:

• **Radioactive pollution**

The radioactivity in drinking water and food products, for the most part, were considerably lower than the regulatory limits set by the Soviet authorities. Occasionally the activity in some food products was found to be above this limit.

• **Radiation doses**

Estimates made by the Committee concluded that the extra lifetime equivalent doses in the most polluted regions would be 80–160 mSv. The Soviet authorities had a higher estimate of 150–400 mSv.

• **Health situation in the polluted regions**

The only serious health effect found among members of the public due to radiation was an increase in the incidence of thyroid cancer (see below).

A number of local clinical examinations, which were not carefully administered, have given confusing and contradictory results. It is evident that some people in the polluted areas were suspicious and believed that they had sicknesses due to radiation even though this was unlikely.

Children were usually in good health. Some children had low hemoglobin counts in their blood, but no differences were found between polluted and clean areas.

Some people have claimed that their immune systems were weakened. Since there is no exact parameter for the strength of the immune system, this is difficult to evaluate. For example, if the number of lymphocytes in the circulation is used as a measure for the immune system, the committee found no differences among people living in polluted and clean areas. There is little evidence to indicate that their immune systems had been weakened due to the accident.

The general cancer incidence in the area has increased in the last decade. This tendency had been noted before the accident. Old reports about cancer are incomplete and part of the increase may be due to improved diagnostic methods and more numerous health checks. No clear increase in the incidence of leukemia has been observed. However, there is a clear increase in the incidence of thyroid cancer for children, which is ascribed to exposures to radioactive iodine during the first period after the accident. Fourteen years after the accident the thyroid cancers (altogether about 700 cases) seem to be the most important somatic effect.

Pollution around Chernobyl

The Chernobyl reactor accident resulted in the radioactive pollution (mainly Cs-137) of large areas in Ukraine, Belarus and Russia. The map indicates regions polluted with Cs-137. The gray areas have a pollution of more than 37 kBq/m². Darker areas are more polluted. The border between the three countries is marked with a dashed curve. The so-called 30 km zone is marked by the circle. Inside this zone, there was additional pollution with Sr-90 and plutonium (see page 113).

Courtesy of the International Chernobyl Project. Assessment of Radiological Consequences and Evaluation of Protective Measures, 1991.

The UN committee UNSCEAR has given a rule of thumb for estimating accumulated doses to people living in areas with Cs-137 pollution. Thus, a pollution of 1 kBq/m² will give an accumulated extra lifetime dose of 0.16 mSv.

This means that the people living in the gray areas for 50 years after the accident must allow for an extra dose above natural background.

Altogether about 825,000 people are living in areas with a Cs-137 pollution of more than 185 kBq/m² (i.e. 5 Ci/km²).

These people may expect an increased lifetime equivalent dose (over 50 years) of about 30 mSv – and more in the most polluted areas. According to the rule of thumb a pollution of about 30 Ci/km² (or 1,110 kBq/m²) would yield an extra dose equivalent to the accumulated natural background dose for 50 years.

It is necessary to take the natural dose into consideration when the biological effects of the Chernobyl accident are discussed.

Radiation-induced cancer has a long latent period. The onset of leukemia, for example, does not reach a maximum until 5 to 7 years after exposure. For solid tumors, the latent period is generally longer. Consequently, it is still too early to arrive at firm conclusions regarding the increased incidence of cancer due to the accident.

So far there is no statistical evidence that the radiation exposure has resulted in damage to fetuses. According to current knowledge about radiation damage, it may be difficult to observe cancer and genetic damage using epidemiological methods (see Chapter 11).

Chernobyl Conclusions

Approximately 14 years after the accident we can make the following observations:

1. The Chernobyl accident was the largest and most severe reactor accident ever. The accident itself resulted in 31 acute casulties, 28 due to the acute radiation syndrome.

2. Large areas were contaminated. People in the regions must live with Cs-137 and Sr-90 contamination for hundreds of years to come.

3. An increase of childhood thyroid cancer has been observed in the most contaminated areas in Belarus, Ukraine and Russia. Since we have used I-131 for medical purposes without similar carcinogenic effect, it is a challenge to extract more information about the radiation doses involved. Furthermore, it is a challenge to understand other factors (biological and environmental) which may influence the risk for radiation-induced cancer.

4. There is no evidence for other radiation-induced cancers in the three most contaminated countries at this time. This is also a puzzle since, according to the doses delivered and the risk models, approximately 500 additional leukemias were expected. Further studies on selected populations, such as the liquidators, are yielding more information. Since a large number of liquidators worked for a longer time in the vicinity of the contamination, we may obtain information about dose protraction, type of radiation and radiation-induced illnesses.

One goal in all these studies is to determine the radiation dose to the people exposed. A possible approach is to study stable chromosome abnormalities such as translocations in the lymphocytes taken from exposed individuals. The FISH technique (Fluorescence *In Situ* Hybridization) offers a new way of examining these abnormalities.

5. Psychological effects and mental disorders seem to be the most severe effect of the Chernobyl reactor accident. It is a fact that a large amount of Post-Traumatic Stress Disorders (PTSD) have appeared with symptoms such as depression, hypochondriasis, headache, dizziness, fatigue or chronic tiredness, poor concentration, anxiety, physical and mental exhaustion, feeling of hopelessness, and lack of libido.

6. For those exposed and/or living in contaminated areas, it has also been observed that there is an increased incidence of high blood pressure, alcohol abuse and even suicide. None of these syndromes are caused directly by radiation.

A consequence of the Chernobyl accident is that millions of people now suffer from psychological effects. The accident has resulted in an increase in what is known as "radiophobia". This needs to be taken seriously. An understanding of radiation and radioactivity combined with the dissemination of properly acquired data will help reduce radiophobia. An important objective of this book is to increase understanding and provide some of the relevant data.

We are of the opinion that knowledge about radioactivity, how to calculate radiation doses, and how to compare doses from accidents with doses from natural radiation, medical use and air-travel is of considerable value to the public. Those who exaggerate the fear of radiation need to take responsibility for increasing radiophobia and the damage spawned by radiophobia.

Plutonium around Chernobyl

This is a map of the region around Chernobyl. The reactor itself is in the middle of the circle, which marks the 30 km zone. The reactor is in Ukraine and the border between Ukraine and Belarus (shown by a heavy curve) goes through the 30 km zone.

The dark area on the map indicates plutonium pollution to an extent of more than 3.7 kBq/m² (0.1 Ci/km²).

The Sr-90 pollution is mainly restricted to the same areas, but extends beyond the 30 km zone. The pollution in this zone is from 37 to more than 111 kBq/m². Doses from these isotopes would be negligible.

$$1 \ Ci/km^2 = 37 \ kBq/m^2$$

(Courtesy of the International Chernobyl Project. Assessment of Radiological Consequences and Evaluation of Protective Measures, 1991.)

Courtesy of The Norwegian Department of Agriculture

The Cs-137 content in a living sheep is measured. Measurements on living animals or humans are based on the γ-radiation emitted (energy of 0.662 MeV). As mentioned above, approximately half of the γ-radiation emanates from a human or animal and can be recorded by an external detector.

Other Reactor Accidents

There were two other major reactor accidents before the one at Chernobyl. However, neither resulted in a significant release of radioactivity.

- **Windscale**

In October 1957, a fire started in one of the graphite moderated reactors in Windscale, England (called Sellafield today). As reported by Crick and Linsley in 1984, the accident resulted in the release of about 600 TBq I-131, 45 TBq Cs-137 and 0.2 TBq Sr-90. The relatively large release of iodine caused some concern and, the day after the accident, I-131 was found in milk. The medical research council suggested that all milk with an activity above 3,700 Bq/l should not be used. This restriction affected the milk from an area of about 500 km². The highest activity of 50,000 Bq/l was found in milk from a farm about 15 km from the reactor. The iodine uptake by the thyroid gland was monitored and the highest thyroid dose was calculated to be 160 mGy.

- **Three Mile Island**

A later and well-publicized accident happened on Three Mile Island near Harrisburg, Pennsylvania, March 28, 1979. The cooling on a pressurized water reactor (PWR) was lost, and parts of the reactor core melted down in the course of 6 to 7 hours before the reactor was covered with water. The reactor had a safety container and only minor amounts of radioactivity were released. In fact, the activity released was smaller than that normally released every year from the natural radioactive sources in Badgastein, Austria, a source that some years ago was considered to be healthy.

Because of some misunderstanding between the Nuclear Regulatory Commission and the authorities, it was recommended that children and pregnant women, living within 8 km from the reactor be evacuated. This recommendation, which was quite unnecessary, had the unfortunate consequence of raising anxiety and fear among the public.

Chapter 10

Radiation and Health. Large Doses

It was noted earlier that radiation can produce biological effects. Around the turn of the century a number of experiments were carried out, which, would be characterized today as dangerous and foolhardy. It was found that ionizing radiation was capable of developing skin burns and could cause hair to fall out. In 1899, Stenbeck and Sjögren from Sweden used radiation to remove a tumor from the nose of a patient. This showed that radiation in large doses could be used to kill cancer cells.

In the early years, when radium was used for treatment, the sources were made in the form of capsules or small tubes. The procedure for radium treatment was either to use a large source of radium (teletherapy) or to use a number of small sources for brachytherapy. In the latter case, paraffin wax was often used and formed to suit the part of the body to be irradiated. Small needles of radium were then sealed into the wax. This procedure gave a good dose distribution when skin ailments and tumors were treated. The general view at that time was that the radiation from radioactive sources was healthy and was a good treatment for most sufferings. Figure 10.1 (next page) shows an advertisement from 1913. Some people made good money by producing radioactive drinking water. A number of small towns in middle Europe such as Badgastein, Baden-Baden, Marienbad, and Karlovy Vary had radioactivity in their water and were considered to be healthy places.

With the use of a jar and some radium salt (Figure 10.1), the water was saturated with radon when radium disintegrated. The belief was that by drinking this water you received "curative" radioactivity. In those days, like today, some

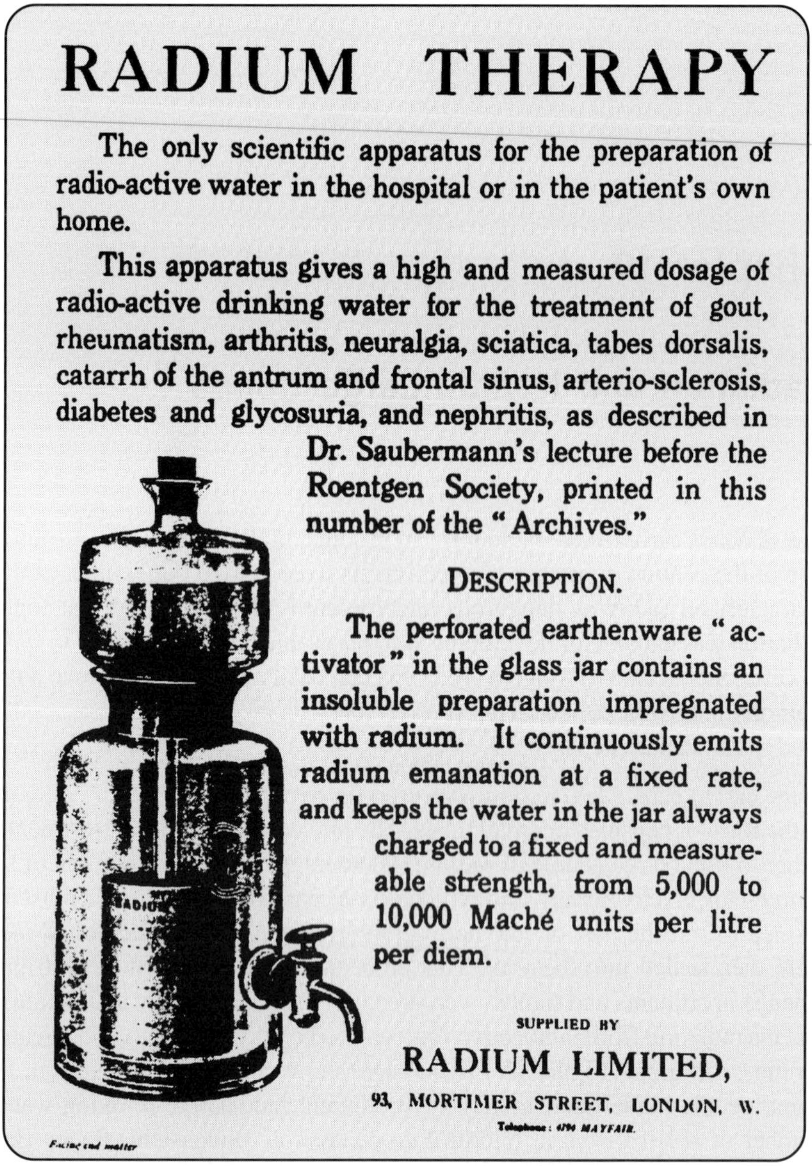

Figure 10.1. An advertisement from 1913 for radioactive drinking water. A jar with water, a cylinder and some radium salt was used. When Ra-226 disintegrates, radon is formed and is released into the water. When the tap was opened, radon was found in the water. The radiation doses were small and the whole system was rather harmless. (Reproduced with permission from R.F. Mould (1980).)

people voluntarily tried out methods that had not been tested or proved effective.

In the early years, people were careless in the use of radiation and the handling of radioactive sources. The reason for this negligence was a lack of knowledge about radiation and its biological effects.

Today, there is great deal of information about the effects of ionizing radiation, a great contrast to the lack of knowledge about the many chemicals in use. However, researchers in the radiation field have not been able to transmit this information to the public. The result is that the public has only an incomplete knowledge about radiation and radiation health effects. In spite of the fact that other human activities are far more hazardous than radiation, many people are unnecessarily afraid.

Because large doses of radiation are known to kill cells, there is the possibilty of using radiation to treat cancer when localized to a small area of the body. Similar large whole-body doses can lead to death, which occurs in the course of days or weeks.

When considering medium and small doses, the biological effect is considerably more difficult to predict. The reason for this is the time lag between the exposure and the observable biological effect. For solid cancers it may be several decades. Marie Curie, and a number of the other radiation pioneers died from cancer; thus, there are reasons to believe that their work with radiation was involved. On the other hand, recent experiments have claimed that small doses may even have a positive health effect (see Chapters 11 and 12).

In all discussions on the biological effects of radiation, the radiation dose is a key issue. Without knowledge about the size of the dose it is meaningless to discuss the effects. The relationship between the dose and effect is also a hot issue in the community of research scientists. Knowledge about the dose–effect curve is a requirement when discussing mechanisms and health risks of radiation.

Dose–Effect Curves

The effect of radiation depends upon the dose. The larger the dose, the larger the effect. This relationship is called the *dose–effect curve* and may be demonstrated with a simple example.

When using an ordinary camera, you know that it is important that the film be exposed to the right amount of light. When exposed to a lot of light, the film becomes black, and when exposed to very little light, the film is hardly darkened

The blackening of the film depends on the amount of light (i.e. the dose).This is illustrated in Figure 10.2. The S-shaped curve obtained is called the *dose–effect curve.*

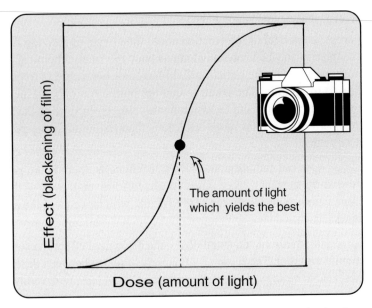

Figure 10.2. Dose–effect curve for the darkening of film. The horizontal axis represents the amount of light (dose) and the vertical axis gives the darkening (effect).

In work with radiation and biological effects, the results are often given by such dose–effect curves. In radiobiology a lot of interest is concentrated on the form of dose–effect curves and, in particular, the form of the curves at small doses. Small doses will be discussed in the next chapter; here the discussion will deal with the effect of large doses.

What is a Large Dose?

In a discussion about the biological or health effects of radiation, the equivalent dose unit, the sievert, is often used. Again, the equivalent dose (Sv) is equal to the physical dose (Gy) multiplied by a radiation weighting factor (w_R). In the case of x- and γ-rays the weighting factor is 1. The dose in gray and the equivalent dose in sievert have the same value. However, in the case of radon and its daughter products which emit α-particles, a radiation weighting factor of 20 is used. Neutrons and other high energy particles have weighting factors larger than 1.

The large dose region can be characterized in the following way:

> A dose of more than 1 to 2 Gy is considered to be large and a dose smaller than 0.1 to 0.2 Gy (100 to 200 mGy) is considered to be small.

Annual doses of 2 mGy to 5 mGy (such as those attained from natural background radiation) are considered to be very small.

The Use of Large Doses

Large radiation doses are used for:

- **Sterilization of medical equipment**
Co-60 and Cs-137 γ-radiation are used to sterilize medical equipment. The doses delivered are on the order of 20 to 40 kGy. The purpose of the radiation is to kill bacteria, viruses, and fungi that contaminate equipment.

- **Radiation of food products**
The purpose is almost the same as for sterilizing equipment. The doses are, however, smaller, on the order of 5 to 10 kGy. Larger doses may change the taste of certain foods.

- **Radiation therapy**
In the radiation treatment of cancer, the purpose is to kill the cancer cells while allowing nearby healthy cells to survive. Much effort is carried out to achieve treatment protocols that will give the most effective treatments. The total dose given to a tumor is 10 to 80 Gy. A treatment protocol may include daily doses of 2 Gy, given 5 days per week. The type of radiation is usually in the form of high energy x-rays from linear accelerators (energies up to 30 MeV) which yield suitable depth dose curves (see Figure 6.3). It is also possible to use electron irradiation but the dosimetry becomes more complicated. The treatment is more effective when the radiation dose is split up into a number of smaller doses rather than giving the same total dose all at once.

The "fractioned dose treatment protocol" has the advantage of partly solving the problems with hypoxic cells in a tumor (see Chapter 12). Experience has shown that fractionated dose treatment yields the best results for tumor destruction while minimizing damage to healthy tissue.

- **Bone marrow transplantation**

In combination with bone marrow transplantation which is used for the treatment of certain illnesses, whole-body radiation is sometimes used with chemotherapy to deplete the original bone marrow. The dose used is about 12 Gy (6 days with a daily dose of 2 Gy). This dose is sufficient to completely destroy the bone marrow and would kill the patient if it were not for the immediate transplant of new compatible bone marrow. A number of people have been treated in this way.

LD$_{50}$ Dose

By definition, an LD$_{50}$ dose (abbreviation for "Lethal Dose to 50 percent") is the dose that could kill 50 percent of the individuals exposed within 30 days. To arrive at a determination of the LD$_{50}$ dose, experiments like the following must be carried out.

In typical experiments, rats, about 15 animals in each group, were given different whole-body doses. The number of animals dying in the course of 30 days was observed for each group. The result is given in Figure 10.3.

The dose is given along the horizontal axis and the number of animals dying (in percent for each group) is given along the vertical axis. The results show that no animals survived a dose of 10 Gy, whereas all rats survived a dose of 5 Gy. It can be seen that the LD$_{50}$ dose is approximately 7.5 Gy.

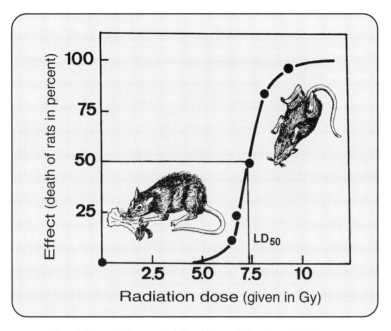

Adapted from A.P. Casarett (1968) with permission from A.P. Casarett

Figure 10.3. Dose–effect curve for radiation-induced death in rats.

When humans and animals are irradiated, the blood-forming organs (in the bone marrow) will be the first to react. For doses of the order 1 to 2 Gy the number of white and red blood cells will decrease as shown in Figure 10.4. As a result of this, the immune system will fail and, after one to two weeks, life threatening infections may occur. If the radiation doses are smaller than 4 to 5 Gy, there is a good chance the bone marrow will recover and resume the production of blood cells. This takes place after 3 to 4 weeks and, consequently, 30 days is a reasonably chosen limit for the name *acute radiation death*. The LD_{50} doses for a number of animals have been determined and some values are given in Table 10.1. Single cell organisms (for example bacteria, paramecium, etc.) may survive doses of the order 2,000 to 3,000 Gy. (This is taken into consideration in radiation treatment of food).

In the case of humans there is not enough information to determine a precise LD_{50} dose. The only information available has come from radiation accidents and the lethality depends not only on the dose and dose rate but also the post-exposure treatment given to the victims.

Table 10.1: LD$_{50}$ doses

Type of animal	Dose in Gy
Dog	3.5
Monkey	6
Rat	7.5
Frog	7
Rabbit	8
Tortoise	15
Goldfish	23
Human	3 - 5

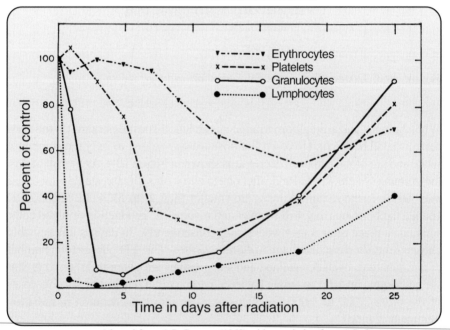

Adapted from A.P. Casarett (1968) with permission from A.P. Casarett

Figure 10.4. Whole-body irradiation results in changes in the number of blood cells. This figure shows the results of a moderate dose to rats.

Acute Radiation Sickness

In 1906, Bergonie and Tribondeau found that there were different radiation sensitivities for different types of mammalian cells. Cells which grow rapidly (high mitotic rate), as well as undifferentiated cells, are the most sensitive. This implies that bone marrow, testes, ovaries and epithelial tissue are more sensitive than liver, kidney, muscles, brain and bone. Knowledge about this is of great importance for those exposed to ionizing radiation. The bone marrow and the epithelial cells of the intestine and the stomach as well as the gonads, the lymphocytes and skin develop the greatest damage. Damage to the bone marrow is the cause of death for whole-body doses in the region 3 to 10 Gy, whereas damage to the epithelial cells of the stomach and intestine is the cause of death for doses in the range from 10 to 100 Gy. For large doses, above 100 Gy, damage to the central nervous system causes death.

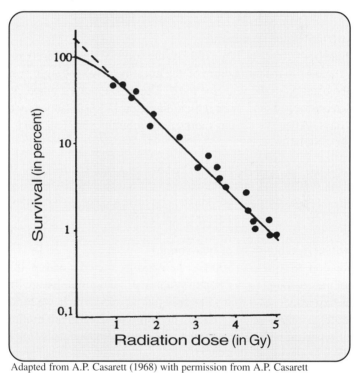

Adapted from A.P. Casarett (1968) with permission from A.P. Casarett

Figure 10.5. Survival curves for bone marrow cells of the mouse after irradiation with Co-60 γ-radiation.

- **Hematopoietic syndrome**

As mentioned above, the failure of the bone marrow is the cause of death for whole-body doses in the range of 3 to 10 Gy. The radiation may either kill these cells or arrest their development. A dose of 5 Gy will kill about 99% of the hematopoietic stem cells in mice (see Figure 10.5). These stem cells are necessary for the production of circulating blood cells (erythrocytes, granulocytes and platelets). A reduction of these cells will result in anemia, bleeding and infections.

The first sign of such radiation sickness is nausea, vomiting and diarrhea. This situation may disappear after a couple of days. Then, the consequences of lost blood cells become evident. Again, significant diarrhea may take place, often bloody, and a fluid imbalance may occur. This, together with bleeding, occurs in all organs. In addition, if infections occur, death may take place in the course of a few weeks.

- **Gastrointestinal syndrome**

For whole body doses of 10 to 100 Gy, the survival time is rarely more than one week. Damage to the epithelium of the intestine results in significant infections from the bacteria in the intestine itself. The production of blood cells is almost completely stopped, and those remaining in the blood disappear in the course of a few days. After 2 to 3 days almost all granulocytes will have disappeared from the circulation.

The symptoms are pain in the stomach and intestine, nausea, vomiting and increasing diarrhea. A considerable loss of liquids and electrolytes will change the blood serum composition. There is an increased chance of infections.

- **Central nervous system syndrome**

For radiation doses above 100 Gy, the majority may die within 48 hours as the result of the central nervous system syndrome. The symptoms are irritability and hyperactive responses (almost like epileptic attacks) which are followed rapidly by fatigue, vomiting and diarrhea. The ability to coordinate motion is lost and shivering occurs followed by coma. Then respiratory problems occur which eventually lead to death.

The symptoms described are due to damage to the brain, nerve cells and blood vessels. Immediately, permeability changes take place in the blood vessels resulting in changes in the electrolyte balance. The loss of liquid from the blood vessels leads to increased pressure in the brain. It is possible that the respiration center in the brain is particularly damaged. Autopsies have shown that some animals die without visible damage to the brain.

A radiation accident

In September 1982 a fatal radiation accident occured in a laboratory for radiation-induced sterilization of medical equipment in Norway. An employee was exposed to a large γ-dose. He was the only person at work when the accident happened. A coincidence of technical failures with a safety lock and an alarm light, together with neglect of the safety routines, resulted in the fact that he entered the room with the source in the exposure position. The drawing below shows the radiation facility. The source is Co-60 with an activity of 2430 TBq.

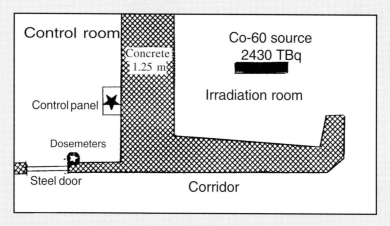

The employee was found outside the laboratory in the early morning with clear signs of illness. Since he had heart problems (angina pectoris), it was first assumed that he had a heart attack. However, it became clear that he had been exposed to radiation. The man had acute radiation syndrome with damage to the blood forming tissue.

His blood counts went down almost like that shown in Figure 10.4. He was treated with antibiotics and several blood transfusions but died 13 days after the accident.

It is important to the know the dose he received. Using electron spin resonance (ESR)-dosimetry the dose was determined (see next page). The man had been exposed to a whole-body dose of approximately 22.5 Gy. The bone marrow dose was 21 Gy, whereas the dose to the brain was calculated to be 14 Gy.

Dose determination and radiation accidents

It is a great challenge to determine the doses in radiation accidents. The reason is obvious because accidents take place without warning and mainly without adequate equipment for dose determination.

Radiation workers usually have a dosimeter and, in the accident described above, the employees used film dosimeters. The film can be used for doses up to about 1 Gy but is not applicable for fatal or near fatal doses. Below, we demonstrate how the radiation dose was determined for this accident.

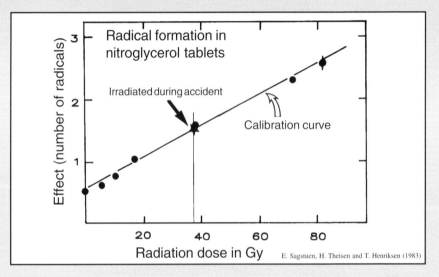

E. Sagstuen, H. Theisen and T. Henriksen (1983)

In order to determine the dose *after* an accident, one must use changes or damages produced by radiation that are relatively permanent. One type of damage that fills this requirement is the formation of free radicals in solid matter (clothes, nails, teeth, etc.).

In the accident described above, the victim used nitroglycerol tablets because of heart problems. He always carried a small box containing these tablets. The tablets were irradiated during the accident and rather stable radicals were formed. The number of radicals formed yields information about the radiation dose. Radical concentrations were measured using ESR spectroscopy (see Chapter 12).

The accidental dose was determined by using the same type of tablets. Exposing them to known doses from 1 to 80 Gy, a calibration curve was obtained. The curve was then used to show that the tablets irradiated during the accident received a dose of about 40 Gy.

Then, measurements were made with a phantom which was placed in the exposure room at the same position as the victim. A number of TLD-dosimeters were used and it was found that a dose of 40 Gy to the position of the box with the nitroglycerol tablets (in his pocket) yielded an average whole body dose of 22.5 Gy.

Conclusion: ESR spectroscopy can be used for dose determinations, even after an accidental exposure.

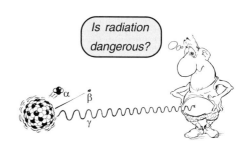

Chapter 11

Small Doses and Risk Estimates

The Dose–Effect Curve

Much is known about the biological effects of large doses of radiation but less is known about the effects of small doses. In most experiments with cells, plants and animals, large doses have been applied with clear and significant results. When the doses become smaller the effects decrease and become less clear. In order to compensate for this, the number of subjects (e.g., animals) can be increased. However, for the region where very small doses are involved (e.g., from an annual dose of a few mGy up to an acute dose of about 50 mGy), the number of animals or humans must be so large that it is very difficult (usually impossible) to conduct experiments and/or epidemiological studies. In epidemiological studies, attempts are made to correlate the radiation dose to the incidence of biological effects such as cancers in a large group of people. Some examples are the populations that have been exposed to radon, those exposed to the bombs at Hiroshima – Nagasaki, and those exposed during the Chernobyl accident. Such studies have yielded both conflicting and confusing results. They are, however, of considerable interest to scientists and to the public.

In this chapter we will discuss known health effects and risks in the low dose region. We will concentrate on the incidence of cancer. The crucial factor is the dose–effect curve.

> The form of the dose–effect curve is essential for all risk estimates.

A discussion of the dose–effect curve for the low dose region must include both experimental work and theoretical models.

The shape of the dose–effect curve must be known in order to evaluate the effects of small increments of dose. Consequently, all risk estimates are closely linked to the assumed shape of this curve. Two very important alternatives are outlined in Figure 11.1 and are discussed in this chapter.

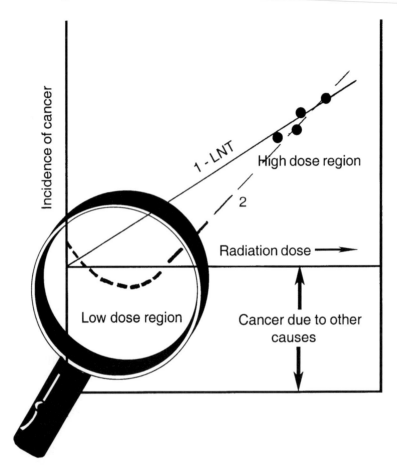

Figure 11.1. Two different dose–effect curves for the incidence of cancer. The curve marked 1-LNT is the well known "linear no-threshold model" for radiation damage. This curve is widely used by the radiation protection community (e.g., the ICRP). The curve marked 2 has an alternative form, including both a threshold (even two) as well as a "hormetic" part for the smallest doses. The filled circles indicate observed data for the large dose region. The two alternatives are drawn to fit observations in the high dose region.

> Risk estimates are based on the form of the dose–effect curve.
> By definition, the risk factor is connected to the steepness (the
> derivative) of the dose–effect curve.
>
> Total risk = risk factor · dose

If the linear no-threshold model (a simple straight line, marked 1-LNT in Figure 11.1) were correct it would be easy to calculate the health effects of radioactive pollution and nuclear accidents like Chernobyl. For a straight line the risk factor is independent of dose. Furthermore, we can use the idea of *collective doses* (see Chapter 4). The calculated total effect using the LNT-model is the product of the risk factor and the collective dose. Such simple calculations have been extensively used and have attracted the interest of the public. For all other forms of the dose–effect curve, however, risk calculations are far more complicated and, for the most part, are impossible.

Small radiation doses never give observable *acute* effects. It is late effects that are observed, consisting mainly of cancer and, to a smaller extent, genetic effects. In our discussion about the form of the dose–effect curve, we include some new research data on repair processes and experiments indicating that small radiation doses may stimulate the immune system and cell growth.

How Can We Get the Needed Information?

Information about the dose–effect curve is obtained in two ways:

1. Experimental. These are methods that utilize data from studies of irradiated animals and epidemiological studies on human cohorts that have been exposed to radiation.

2. Theoretical. These methods utilize models based on the mechanisms for carcinogenesis. If we assume that radiation-induced cancer is a stochastic process, determined only by random ionizations, the dose–effect curve would be linear. If, however, other dose-dependent processes are initiated that interact with the cancer forming processes then a linear response can no longer be assumed.

Experimental Information on Radiation and Cancer

It may seem strange to many that radiation which is used for cancer treatment also represents a risk for the formation of cancer. This duality is due to the fact that radiation may initiate several processes in the cell. For large doses the cells may be killed and a dead cell can never develop into a cancer cell. However, if the cell is only damaged, and the damage not repaired or misrepaired, the cell may subsequently transform into a cancer cell which in turn can divide several times and form a tumor.

A few years after the discovery of radiation, exposed people developed cancers. A number of medical doctors, using x-rays daily, developed *squamous cell carcinomas* on their hands and arms, radium dial painters got bone cancer and the miners from the Hartz area in Germany got lung cancer. Madame Curie died from cancer which probably was caused by her work with radiation.

Before we embark on observations which can yield information on the dose–effect curve we shall discuss a few important topics:

- **Type of cancer**

The majority of cancer types are induced by radiation. Certain types, such as leukemia and thyroid cancer, appear to be more frequent than others. It is, however, important to stress that since cancer is a rather common disease, radiation generally plays a *minor* role in causing this illness. Therefore, it is rather difficult to decide whether a particular cancer incidence in a population is caused by radiation or if other cancer causing factors are involved.

> No particular type of cancer is formed by radiation

- **Latent period**

In the case of cancer, whether induced by chemicals, smoking, or radiation, it is known that there is a lapse of time between the exposure and the time when the diagnosis of cancer is made. This period of time is called *the latent period*. There is very little information about what takes place in this period but several mechanisms have been proposed for the development of cancer.

One mechanism for cancer induction is the possibility that a cell damaged by radiation is mutated. If the damage consists of a genetic change, that is either not repaired or is misrepaired, it is called *a somatic mutation*. A similar change in a sex cell (gamete) is a *genetic mutation*.

Mutations are caused by damage to DNA or damage occurring during the cellular division processes. These damages result in a transformation of the healthy cell into a cancer cell. If, or when, this primary cancer cell divides, two cancer cells are formed. In order to observe a tumor of the size of a pea, 20 to 25 cycles of cell divisions must take place. The time elapsed between the damaging event and the detection of cancer is the latent period (see Figure 11.2).

The latent period can vary from a few months up to a number of years. The incidence of leukemia after the bombings in Japan reached a maximum after 5 to 7 years. In the Chernobyl accident, thyroid cancers among children initially

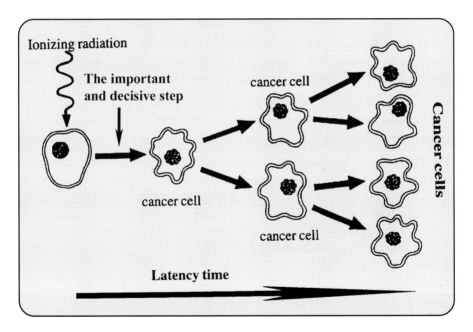

Figure 11.2. The mutation theory for cancer hypothesizes that a normal cell is transformed into a cancer cell. After one cell division there are two cancer cells. The number increases from 2 to 4 – 8 – 16 – 32 and so on. After 20 cell cycles there are more than one million cells and the tumor may be diagnosed. The time elapsed between radiation exposure and detectable cancer is known as the latent period.

appeared 6 to 7 years after the accident. For other cancer forms, the latent period may be as long as 10 to 20 years or more.

The fact that the risk is associated with a latent period is considered unusual by many, even when considering lung cancers from smoking. In the case of traffic accidents, the damage usually takes place immediately. On the other hand, late effects may also occur with traffic accidents.

It should also be mentioned that the latent period seems to depend on the dose. Longer latent periods are observed for smaller doses. In this connection, it should be noted that *if the latent period becomes very long, extending through the rest of a person's expected lifetime, the cancer will never appear.*

* **Dose threshold**

One of the key issues that will be discussed in this chapter is whether a radiation dose threshold exists. The existence of a threshold means that below a certain dose there is no risk that cancer will be induced by that dose.

As we shall see, neither experimental animal data nor epidemiological human data have, so far, solved the problem as to whether or not a radiation threshold exists.

If we look to theoretical models, it appears that the stochastic theory, starting out with a single hit (i.e. an ionization), would imply no threshold. A number of people within the radiation protection community claim that it can not be excluded that a single ionization or a single track of ionizations results in a cancer. It is difficult to test this possibility; as shall be seen, radiation is an agent that influences more than one process in the living cell. While some are negative, others may be positive.

With regard to the one-ionization or one-track theory, we would like to point out that a very large number of ionizations (and tracks) are produced in our bodies (in effect continuously) because of natural background radiation. As you can see from the calculations on the next page, the radiation from natural sources yields approximately 500 million ionizations in an adult per second!

> *It may well be that we will never gain enough information to conclude whether a threshold dose exists for radiation-induced cancer.*

Ionizations in the body from natural radiation

Most people are surprised and somewhat skeptical when physicists say that the natural background radiation results in about 500 million ionizations in the body *per second.* If you want to confirm this read the following.

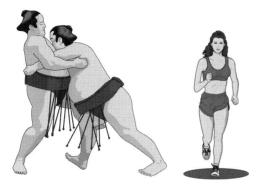

The number of ionizations is proportional to the body weight. For a sumo wrestler it can easily reach 1 billion per second, whereas a jogging woman hardly reaches half of that.

Figure 7.1 presents the different radiation sources and the annual equivalent doses to an average person. The equivalent doses are given in mSv whereas in the present calculation the dose unit Gy must be used. This means that the radon dose, which entails using a large radiation weighting factor (because of the α-particles) will be reduced considerably. The natural sources yield an annual dose of approximately 1.5 mGy. This means that 1.5 mJ (millijoule) of energy is absorbed per kilogram in the body, or $9.4 \cdot 10^{15}$ eV.

$$1 \text{ eV} = 1.6 \cdot 10^{-19} \text{ J.}$$

The energy absorbed results in the formation of ions and excited molecules. The average energy used to produce an ionization in air is known to be 34 eV. Let us assume that 34 eV also will produce an ionization in our bodies. From this, the total number of ionizations per kilogram per year may be calculated. The result is $2.75 \cdot 10^{14}$ ionizations per kilogram per year.

In a person weighing 70 kilograms, the number of ionizations in the body per second (N) is given by:

$$N = \frac{2.75 \cdot 10^{14} \cdot 70}{60 \cdot 60 \cdot 24 \cdot 365} = 610 \cdot 10^{6}$$

This very simple calculation demonstrates that the number of ionizations in our bodies is about half a billion per second. This is a very large number and nothing can be done about it. In employing the linear no-threshold theory, any of these ionizations, or at least a cluster of ionizations within a track, may be the crucial one for the biological damage.

- **Dose rate**

The dose rate is, by definition, dose delivered per unit time, i.e. how fast the dose is given. Dose rate is an important factor in radiation biology and, more specifically, in the induction of cancer. A certain dose given within a short time interval has a larger effect than if the same dose is protracted. In risk analyses, a *dose and dose rate effectiveness factor* (DDREF) is introduced.

The concern with the dose rate is connected to repair processes and the cell's adaptation to radiation. If damage is accurately repaired the consequences disappear. In the case of a high dose rate, a large amount of damage produced in a short period of time may overwhelm the repair systems, which can only work so fast. It is reasonable to assume that the fraction repaired under such circumstances is smaller than that obtained when the repair system has more time available, e.g. at a low dose rate. Another interesting behavior is that cells seem to be adapted to or stimulated by small amounts of radiation. This raises the question whether a certain amount of chronic radiation is necessary for a healthy life. So far, no cells or organisms on earth have lived without ionizing radiation.

Radiation-induced Cancer in Animals

A great deal of our knowledge about radiation-induced cancers is from experiments on rats and mice. The animals have usually been irradiated with rather large doses (more than 0.5 Gy) and the number of cancers induced have been observed. The experiments involve both whole-body irradiation as well as local irradiation with doses to certain organs. Examples from a couple of experiments, using whole body x-radiation, are shown in Figure 11.3.

The upper curve in Figure 11.3 shows a dose–effect relation which reaches a maximum and then goes down for doses above 2 to 3 Gy. This response does not lead to the conclusion that the risk decreases for large doses. The fact is that for doses in this region some of the animals die (from acute radiation syndrome) and can, therefore, not get cancer. The two alternative effects were mentioned earlier: transformation of cells and killing of cells. For large doses, the killing effect dominates for both cells and animals.

In the lower curve, the results are given for an experiment in which a group of mice were exposed and the formation of cancer in the ovaries observed. The radiation was γ-rays from a Cs-137 source. In this experiment, it was shown that the dose rate was important. The particular curve given is for a dose rate of 83 mGy per minute. When the dose rate was increased to 450 mGy per minute the cancer incidence increased considerably. The figure shows that doses below 1 Gy yield few cases of ovarian cancer. Other animal experiments are in line with these results.

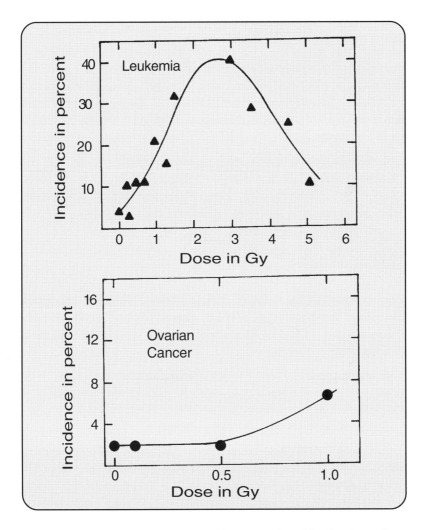

Figure 11.3. Dose–effect curves for radiation-induced leukemia and cancer in mice. The upper curve is for leukemia and the lower curve for ovarian cancer. In both cases, the doses are much larger than the small doses discussed in this chapter.

> *Altogether, the experiments with mice yield information about radiation and cancer for large doses but little information with regard to small doses.*

Epidemiological Studies

A number of people have received considerably larger doses than average, either at work or otherwise. These cohorts can be studied for the incidence of cancer. Such studies may yield information on the relation between cancer incidence and the radiation dose. The goal is to get a measure of the risk of cancer at low doses.

The task described above has two important parameters: one *medical,* in which recording the onset of disease is important (e.g. the diagnosis of cancer), and one *physical,* which consists of an accurate determination of the radiation dose. The latter parameter is by far the more difficult one to ascertain.

The radiation dose

It is easy to determine the radiation dose in planned laboratory experiments. In the case of accidents, however, it is far more difficult. People do not always carry dosimeters with them and scientists are left to calculate the dose from information attained after the exposure.

In the case of protracted doses, i.e. when the extra dose is received over days, weeks and years, the situation is even more difficult. On the next page an attempt is made to give you an idea of some of the problems encountered in the determination of low doses given over long times.

In a previous chapter we have shown that the natural radiation sources yield chronic radiation with an annual dose of from 2 mSv to more than 10 mSv. However, none of us is an average person. We receive relatively continuous doses all the time interlaced with larger spikes when visiting the dentist, traveling by air, getting an x-ray, etc. Some people also receive extra exposures at work (for example, some medical workers and air crews). Some people are also exposed to extra doses because of accidents (e.g., Chernobyl), either acute or protracted over several years.

This shaded section can be skipped. For the interested reader, it points out some of the problems encountered in finding control groups in epidemiological studies with small extra radiation doses.

Considerations on accumulated doses and epidemiology

The fact that we are exposed all of the time to annual doses ranging from about 2 mSv to more than 10 mSv makes it difficult to ascribe any health effect to additional or extra doses that are within the natural variation. The natural sources (radon, external γ-radiation, internal radiation and cosmic radiation) yield a chronic dose level. Some average values are given in the table below. However, variation within the different areas may be up to a factor 10 (mainly due to radon and external γ-radiation). For example some areas within the region of Yangjiang in China have a high natural background radiation and the population receives annual doses of more than 17 mSv.

World average	2.5 mSv/year
U.S.A.	3.0 mSv/year
U.K.	2.6 mSv/year
Scandinavia	3.2 mSv/year
Yangjiang	6.4 mSv/year

In addition to the exposure from natural sources, we must also include doses from medicine, dentistry and industry. Furthermore, certain groups are also exposed at their workplace.

In the discussion on the effect of small doses, the background dose, particularly the accumulated background dose, is very often neglected. The variation in the background dose from one person to another can be significant, and in Figure 11.4 an attempt is made to describe the problems. Most people do not have a constant annual dose. Some years you probably have several x-rays, or air trips, or you simply move from one house to another within the same area. All these things can drastically change the annual dose and consequently the accumulated dose; i.e. the accumulated dose curve for a person is not a straight line. The normal variation in accumulated dose after 70 years would be in the range of about 150 mSv to more than 1000 mSv.

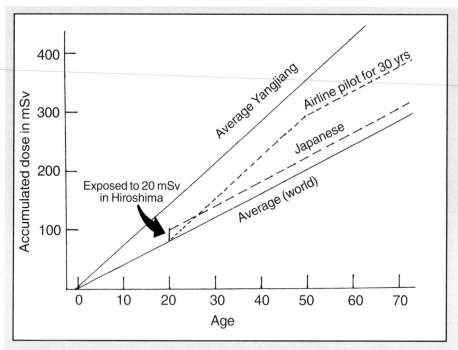

Figure 11.4. The figure depicts in an approximate way how we are exposed to radiation from birth (or rather conception) to grave. The accumulated dose curves are not straight lines for any person. While the solid curves show the average for the entire world and one select population in China, the two dashed curves are for two individuals with different histories. One is Japanese, who at age 20 (in 1945) received an acute dose of 20 mSv from the bomb. The other one is an airline pilot who received an extra annual dose from cosmic radiation of about 3 mSv per year during his 30 year career.

Based on the curves shown in Figure 11.4 the crucial question is: would it be possible to observe any biological effect from a small extra radiation dose (acute or protracted) of 1mSv to 30 mSv? The answers are not available but work carried out with the purpose of observing health effects of small doses *must include the background doses and their variations.*

In Chapter 9 we discussed doses from the bomb tests and the Chernobyl accident. The doses to the majority of people around the world were below 1–2 mSv. For a few exposed groups the accumulated dose was more than 10 mSv. It can, therefore, be concluded that we will hardly observe any biological effect from these extra doses.

Cohort Examples

A number of the pioneers who used radiation died from cancer and it is reasonable to assume that they received large radiation doses. The first groups were radiologists who got skin cancer and uranium miners who suffered from lung cancer.

As noted earlier, the numerals and hands on some old clocks were painted with radium. Women were employed to do the painting. Sometimes they ingested small amounts of radium because they "pointed" their brushes by licking them. Radium is a bone seeker and some of the women contracted bone cancer later in life.

Several years ago when tuberculosis was more frequent, patients were examined with x-rays. A number of x-rays and fluoroscopies were performed in combination with pneumothorax treatment. The idea was to give the lungs some rest which was accomplished by collapsing the lung with air (today antibiotics are used). The treatment could last for several years and the patient needed more air every second week. Each fill of air required more x-rays and the doses could be significant. It has been observed that an increased incidence of breast cancer occurred in the women that went through this treatment.

For a while, a radioactive material (thorotrast) was used as a contrast medium in x-ray diagnoses. Increased cancer incidence (liver cancer) has been observed among patients so treated, particularly in Germany and Japan.

In the groups mentioned above, all received radiation doses larger than average. However, dose determinations were not made at the time and, so far, it has not been possible to arrive at any dose–effect curves which may be used for risk calculations.

Hiroshima and Nagasaki

The people who survived the bombs in Hiroshima and Nagasaki have been used in studying radiation-induced cancer. It is estimated that about 429,000 people lived in the two cities in August, 1945. The bombs killed about 67,000 the first day and about 36,000 in the following four months. In 1950, about 283,000 people claimed that they were exposed to radiation because of the bombs. The average dose is estimated to have been approximately 160 mSv.

In the period from 1950 to 1978, a total of 67,000 of this large cohort died. Approximately 530 cancer deaths can be attributed to the radiation dose from the bombs, 190 leukemia and 340 of other cancer forms. This yields a ratio of 2.79 to 1 between solid tumors and leukemia. In another more recent study on a group of 42,000 with an average dose of 300 mSv, 80 cases of leukemia and 260 other cancers (in the period 1950 to 1985) were found, yielding a ratio of 3.25 to 1. The ratio between solid tumors and leukemia appears, therefore, to increase with time and is indicative of different latent periods for the different types of disease.

In order to arrive at the risk for radiation-induced disease, it is necessary to determine the doses attributable to the bombs. These were acute doses delivered within seconds in August, 1945.

Little information exists about the accumulated doses received, as well as of the variation in doses from one person to another during the subsequent 50 years. It is reasonable to assume that in the period since the war the accumulated equivalent dose is on the order of 100 to 200 mSv. This equivalent dose, and in particular its variation, has been very difficult to take into consideration.

Dose estimations were published in 1966, 1981 and 1986. Two different circumstances makes these calculations difficult.

1. The radiation quality
The radiation from the bombs was a mixture of γ-radiation and neutrons (and neutrons are given a radiation weighting factor of from 5 to 20 depending on the energy).

2. Weather conditions
Neutrons are readily absorbed by water. Consequently, the weather conditions and humidity are of importance for the dose evaluation.

To estimate individual doses, knowledge is needed about the location of the person during the blast and what kind of protection he or she had (being indoors, outdoors, etc.). Some years ago, it was found that the estimated position of the centre of the explosion at Nagasaki had to be changed by 37 meters. This may sound trivial, but the result was that some persons had to be moved from one dose group and placed into another.

The bomb material which envelopes the fissionable material (U-235/Pu-239) is important with regard to the dose. In Hiroshima and Nagasaki iron was used and a heavier metal in the nose and tail. This would slow down some of the neutrons and absorb some of the γ-radiation. Some years ago the assumption was that the neutron doses were large in Hiroshima compared to those in Nagasaki. Later, new calculations showed that the neutron doses in Hiroshima had to be reduced and the γ-doses set higher.

As you can see from these considerations, it is both difficult and tedious to determine the radiation doses received by the people in Hiroshima and Nagasaki in August, 1945. In spite of this, a selected group of survivors have been chosen and followed through all the years from 1945. The studies of this group, the so-called *Life Span Study Cohort*, will continue until about 2015 (a period of 70 years).

The life span study cohort

Most of the work carried out to obtain information on the form of the dose-effect curve is based on the life span study cohort. The acute equivalent doses to these people varied considerably and, according to the determinations from 1986, about 52% received equivalent doses below 10 mSv.

The people within the life span study cohort were grouped together depending on the acute equivalent dose. The dose levels selected were: 0–10 mSv, 10–20 mSv, 20–50 mSv, 50–100 mSv, 100–500 mSv, 500–1000 mSv, 1–2 Sv, 2–3 Sv, 3–4 Sv and above 4 Sv.

The results of this study, as of the mid 1990s, are given in Figure 11.5. The acute equivalent doses are given along the horizontal axes and the relative risks for cancer deaths are given along the vertical axes. The relative risks for the groups 0 to 10 mSv are set equal to 1.0.

As you can see, the accumulated equivalent dose from natural radiation, before and after August 1945, is not included. This is a weak point and makes it difficult to draw any firm conclusions about the dose effect curve for doses below about 200 mSv.

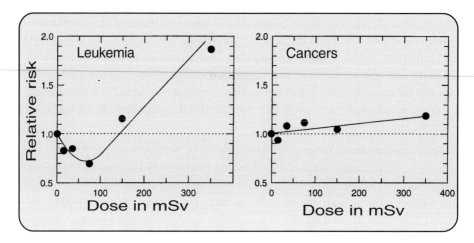

Figure 11.5. The figure indicates relative risk for radiation-induced leukemia and cancer death. The equivalent dose calculations are from 1986 and the data refer to the life span study cohort. Note that the doses given here refer to the extra dose received immediately in 1945.

As seen from Figure 11.5 the results for leukemia are compatible with an u-shaped dose effect curve (see Figure 11.1), implying that small acute equivalent doses may even protect against leukemia. This would be a somewhat hasty conclusion, but demonstrates clearly that, for doses below 100 to 200 mSv (acute doses), sufficient information is lacking about the shape of the dose-effect curve. On the right side of Figure 11.5, it is shown that the low dose data for solid tumors can be fit by a straight line. However, if the high-dose data are included, it appears that no single straight line fits the data for both the low and the high dose region.

Conclusion about the Dose–Effect Curve

We started this chapter with the purpose of arriving at the form of the dose–effect curve and stated that knowledge about this curve is essential for risk estimations of ionizing radiation.

The information attained so far is from laboratory experiments on animals as well as epidemiological studies on different exposed groups of humans. We have mentioned the life span study cohort from the bombs in Hiroshima and Nagasaki. The study of this group has provided information on the cancer risk for large acute doses. Whether these data can be used in the case of small doses and be integrated with an accounting of the accumulated doses (attained in the years following the bomb) is not known.

Other possible cohorts are people living in areas with large background radiation (either from natural sources or due to accidents like Chernobyl). In the case of the Chernobyl accident, it is extremely difficult to determine individual doses. Some of the radiation doses to workers and the public were acute (in particular to the liquidators). The people now living in the polluted areas will attain a larger annual dose than before the accident. The extra dose due to Cs-137 pollution can be reduced by a restrictive food market.

We also note that some other groups have received extra doses related to their profession (radiation workers). For all these cohorts the crucial parameters are the radiation dose received by the cohort and identify a good control group. Finding a control group, a group that is equivalent except for the extra dose, is challenging. For example, it is very difficult to make sure that the control group has received a comparable background dose.

In China, two large groups of people, living in areas where the annual background dose varied by a factor 3 have been compared. It appears that for cancer deaths, the area with the lower dose exhibited the higher risk. This result clearly demonstrates the problems that exist for arriving at the dose–effect curve for small doses.

The conclusions so far are:

> *1. Large doses of radiation may induce cancer.*
>
> *2. We lack sufficient information about the shape of the dose-effect curve for doses below 200 mSv.*
>
> *Consequently, we have to find alternative approaches to solving the low dose problem. One very promising approach is the study of the basic mechanisms by which a normal cell transforms into a cancer cell.*

The dose–effect curve and cancer research

The assumption that cancer is the end result of a stochastic process would result in a linear dose–effect curve for the low dose region. This simple model is questionable and needs to be tested. A very promising procedure is to attain more information about the basic mechanisms involved in carcinogenesis.

In Figure 11.6 we give a glimpse at a few of the key molecular elements that play a role in the mechanisms of carcinogenesis. The basic idea is that a mutation in a critical gene, or genes, can lead to a tumor. Thus, a mutation in a tumor-suppressor gene, such as p53, has been found in many human cancers. The p53 gene codes for the p53 protein, a protein that controls a check-point in the cell cycle. Inactivation of this gene, caused by a mutation, is one factor that permits damaged cells to divide.

A mutation is due to damage in the DNA molecule. In Chapter 12 we describe some of the research on damage and repair of the DNA molecule. It is important to realize, however, that normal metabolic processes in the cell routinely produce large amounts of DNA damage (called endogeneous damage). Radiation and chemicals increase the amount of DNA damage (called exogeneous damage). Repair mechanisms can excise most of this damage, but the amount of unrepaired damage increases with age, perhaps in part due to a decreasing efficacy of the repair system.

The presence of lesions in the cell seems to have a stimulating effect on DNA-repair. This is supported by some recent studies which demonstrate that radiation has an adaptive effect (see Chapter 12.7). Small amounts of radiation seem to increase or stimulate the repair processes as well as the mechanism of apoptosis (programmed cell death). It is, therefore, possible that the effect of both endogeneous and exogeneous damage to the cells is reduced.

> Thus, the possibility must be considered that a small amount of radiation can both stimulate apoptosis and make the repair system more effective than it would be in the absence of the small dose. This would support a U-shape at the low dose end of the dose–effect curve, as depicted in Figure 11.1.

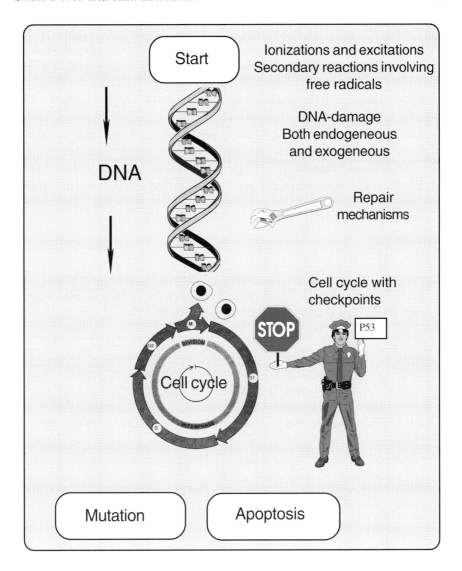

Figure 11.6. An attempt to decribe some of the important steps and products involved in the transformation of a normal cell into a cancer cell. Ionizing radiation can add some extra damage to the DNA. A cell with damage can result in a mutation and a cancer cell. Other alternatives are repair or the cell can be triggered into a programmed cell death (apoptosis). The damaged cell can also be stopped at checkpoints in the cell cycle. New research seems to demonstrate that several of these processes are stimulated by radiation.

A damaged cell may also be stopped in the cell cycle at certain checkpoints. The p53 protein controls a checkpoint in the G_1 phase and can prevent damage from evolving into a mutation. This protein is released in the repair process. The p53 protein is also involved in triggering cell death (apoptosis) which will prevent a damaged cell from being transferred into a cancer cell.

We hope that this brief review provides some insight into the many steps and products that are involved in carcinogensis. To date, an understanding of these mechanisms is just beginning to emerge. It is, however, an interesting and active research field and there is little question that continued research will bring new understandings.

Conclusions

In the previous sections we have summarized some different procedures that can be explored in an attempt to attain more knowledge about the dose-effect curve for small doses. The following conclusions are made:

> A. Neither experiments nor theoretical models are at present available for determination of the dose–effect curve for small doses (i.e. doses in the region up to about 100 mGy).
>
> B. The dose–effect curve is essential for all risk and cost-benefit analyses.

In the next section we shall examine, in more detail, the consequences of using the different dose–effect curves.

Risk analyses

The ICRP, as well as radiation protection authorities in most countries, uses the linear no-threshold model (LNT) which is easy to manage in risk analyses both for planning purposes and after radiation accidents. One attribute of the LNT-model is that those in authority can use the concept of collective dose in the risk calculations.

Collective doses

The collective dose is by definition the sum of the individual doses of all of the exposed people or the average dose multiplied by the number of people (see Chapter 4). It should only be applied in cases where the dose effect curve is known (or assumed) to be linear. For all other forms of the dose-effect curve, the collective dose concept has little meaning; under these circumstances it should not be used in risk analyses.

As discussed above the linear no-threshold model yields a constant and dose-independent risk value for the detrimental effects of radiation (curve marked 1-LNT in Figure 11.1). Only one single point on the dose-effect curve is needed in order to arrive at the risk value (the steepness of the curve). The ICRP has several times tried to give a risk number for radiation induced cancer death. The risk values have varied from $6 \cdot 10^{-3}$ to $50 \cdot 10^{-3}$ per Sv. The latter value is used in the ICRP 60 report.

Let us now give you a few examples of risk analyses based on the linear no-threshold model. It is based on very simple mathematics, involving the product of the collective dose with the risk value ($50 \cdot 10^{-3}$ per Sv).

For the United States with a population of 263 million and an average annual equivalent dose from natural radiation sources of about 3 mSv, the collective annual equivalent dose is around 789,000 person-Sv. Using the above risk factor, one calculates that approximately 40,000 Americans would die from cancer each year due to the background radiation. This is about 10% of the observed annual rate for death by cancer.

Another example is the use of radiation in medicine. A hospital cannot be run without x-rays and the use of x-rays adds an annual equivalent dose that averages out to about 0.6 to 1 mSv. This tiny dose results in a collective dose to the US population of 158,000 Sv to 263,000 Sv. With a risk factor of $50 \cdot 10^{-3}$, this medical dose would result in 8,000 to 13,000 cancer deaths each year.

The LNT model has been used to calculate risk due to the Chernobyl accident.

Since the collective dose is the product of a very small individual dose with a very large number of people the result is a large collective dose. Public fear of radiation is promoted by such analyses. It is important, therefore, to recognize the weakness in such calculations. In order to better understand this, consider the following example:

If 10 people each take 1,000 tablets of aspirin the collective aspirin dose is 10,000 person-tablets. It is a possibility that at least 9 out of the 10 will die. If 10,000 people each take one aspirin tablet the collective dose is still 10,000 person-tablets. It is an extremely low probability that any of the people will die after taking one aspirin tablet. However, if the linear no-threshold dose effect curve is used, the conclusion would be that at least 9 persons will die from a collective dose of 10,000 person-aspirin tablets.

Other forms of the dose–effect curve

For all other forms of the dose–effect curve, the concept of collective dose would be without meaning. Furthermore, the risk factor would depend on the radiation dose. In addition, there remains the possibility that small radiation doses may have a beneficiary effect (the U-shaped curve in Figure 11.1).

Can we live without radiation?

An interesting, but hypothetical question is: do we need a small radiation dose in order to be healthy, or could we improve the public health by moving to an environment free of radiation? In fact, as already pointed out, there is a large variation in doses from the natural radiation sources. There is no evidence, as of yet, that health is improved in regions having a low radiation background. In fact, some results have been published pointing to the contrary.

We have no information how it would be to live without ionizing radiation. We can, however, mention some experiments carried out by Planel and coworkers (1987) on some single cell animals. They studied the growth of *Paramecium* (the small "slipper shaped" cells living in water) by carrying out the following experiments:

A. Cultures of paramecia were put into a 5 to 10 cm thick lead chamber that reduced the background radiation to almost zero. The result was that cell growth was reduced. The same result was obtained when the experiment was carried out in an underground laboratory where the radiation background was very small.

B. The next step in the experiment was the introduction of radioactive sources. The radiation from the sources resulted in a radiation level which yielded an annual dose of from 2 to 7 mGy (comparable to normal background radiation). The result was that the the cell growth increased back to "normal". These and similar experiments indicate that small doses of radiation may stimulate a number of processes such as cell growth and cell repair.

How to treat small doses of radiation?

We started this chapter with the goal of describing a dose–effect curve in the region for small doses, which could then be used for radiation protection and risk estimates. This is a very important task that has large consequences for all of us, mainly for psychological and economic reasons. This goal was not achieved; to date a dose–effect curve for the region from zero to about 100 mSv remains undetermined.

Given the lack of data on low dose effects, the ICRP and radiation protection authorities use the linear no-threshold model. In combination with the concept of collective equivalent dose, it has been used to calculate the detrimental effects of radiation to the public. While application of this model is straight forward, it does not settle the question on how to manage the small extra equivalent doses that may be due to accidents, the nuclear industry and waste disposal. A guide to this is the ALARA principle, which states that the equivalent dose should be kept "As Low As Reasonably Achievable". The word "reasonable" is a key in all cost benefit calculations. Thus, in the linear no-threshold model, the individual doses of 1 mSv can be of significant importance in a large population. This was clearly demonstrated after the Chernobyl accident. In a number of countries in Western Europe, the maximum allowable level set for Cs-137 in food products was exceeded. Large amounts of money were spent in order to reduce the average personal dose That dose, when accumulated over 50 years, would be about 1 mSv. From a radiobiological point of view this is wasted money. Furthermore, restrictions of this kind have contributed to increasing the fear of radiation. That fear is a large burden to society.

We have tried throughout this book to point out that we are living in an "ocean of radiation" which gives us an annual dose of 2 to 4 mSv, on average. In addition, we use radiation for medical purposes, in industry and research, that on average adds another 1 mSv to the annual exposures. How should the society treat this radiation?

The difficulties in obtaining reliable data in the low dose region are likely to persist, as are the difficulties in interpreting epidemiological data. Considerable promise, however, exists in approaching the low dose/low dose rate problem through understanding the basic physics, chemistry, and biology of radiation action. In the next chapter, we turn our attention to studies on the radiobiological mechanisms that are important to understanding radiation induced carcinogenesis.

Small equivalent doses
you may be exposed to

If you go to the dentist, travel in an airplane, consume food containing Cs-137 or take an x-ray picture, you receive small radiation equivalent doses that come in addition to the background radiation that is always present. Here are some examples of the size of these doses.

Reindeer dinner: (1,000 Bq/kg in meat)	~ 0.003 mSv	X-ray CT, fluoroscopy: (stomach and intestine)	~ 1–20 mSv
X-ray (dentist):	~ 0.03 mSv	X-ray (kidney, back):	~ 1–5 mSv
Air travel US-Europe:	~ 0.03 mSv	Annual dose for air crew:	~ 2–3 mSv
X-ray (thorax):	~ 0.1 mSv	In space for a month:	~ 10 mSv

Chapter 12

Radiation Biology — Mechanisms

In this chapter we will discuss some of the basic mechanisms for the action of radiation on biological systems. Under discussion will be how radiation can change or damage the molecules in a cell, how it can influence the cell cycle, and how radiation acts upon animals and humans.

The absorption of radiation energy initiates a large number of processes in a cell. It may take some time (weeks, months and years) before the biological result can be observed. This situation must not lead us to believe that nothing takes place in this latent period.

The starting point is the absorption process (see Figure 12.1). Radiation energy is deposited in the system and a number of "primary" products are formed. These products (ions, excited molecules, and free radicals) are very reactive with lifetimes in an ordinary cell of the order of a fraction of a second. Their reactions with molecules in the cell result in secondary processes which finally yield a macroscopic result such as cell death, cancer or genetic change.

The research fields called "radiation research", "radiation biology" or "radiation sciences" include physics, chemistry, biology and medicine and examine all of the processes shown in Figure 12.1.

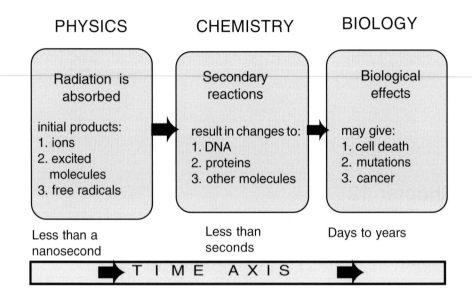

Figure 12.1. The radiation-induced processes in a biological system.

One hundred years of research on the applications of radiation have yielded a large amount of information. Early on it was observed that radiation sensitivity varied from one cell type to another and that the sensitivity was greatest for cells in rapid growth. This observation initiated the use of radiation in cancer treatment since these cells seem to grow rapidly under little control. The field of radiation biology was first promoted in the excellent book of D.E. Lea from 1946 (*Actions of Radiations on Living Cells*).

Because it is not possible to cover all the interesting areas of research in radiation biology, the goal here is to present selected topics that will provide a sense of the scope and the progress being made.

Radiation Biophysics

The effect of any ionizing radiation starts with an interaction between radiation and the molecules in the cell. There are two different types of interactions:

- **Direct effect**
The effect is observed in the same molecule where the primary absorption occurs.

- **Indirect effect**

In this case the radiation energy is absorbed in another molecule (mainly water), resulting in the formation of reactive products that subsequently react with other molecules in the system.

In a dry system (without water), only the direct effect occurs, whereas in an aqueous system the indirect effect dominates. Reactive water radicals are formed that initiate a number of subsequent processes. A living cell consists of about 70% water and 30% other materials. In such a system, the direct and indirect effects are approximately equally important.

Study of the primary radiation products is difficult because the life times of intially formed products are only milliseconds or shorter. There are two main strategies for studying these short-lived reactive products (called unstable intermediates):

1. **Rapid techniques**

The approach is to observe unstable intermediates in a very short time span before they dissappear. In one technique, unstable intermediates are created by a short intense pulse of radiation (less than a nanosecond, 10^{-9} s). The intermediates are then detected by a very rapid measuring system, typically looking at changes in properties such as light absorption, emission, and conductivity. This approach is called pulse radiolysis.

2. **Stabilizing methods**

The lifetime of unstable intermediates can be increased by using low temperatures, effectively slowing down and even stopping reactions. Very often this entails using temperatures below $-100°C$. In order to attain these temperatures, liquid nitrogen with a temperature of $-196°C$ or liquid helium with a temperature of $-269°C$ is used to cool the sample. Experiments with liquid helium are very informative but difficult to do. The intermediates are first stabilized at extremely low temperatures. Then by slowly warming it is possible to observe the reactions as they unfold. This procedure makes it possible to study the secondary reactions.

For a number of years research groups have studied unstable intermediates formed in hormones, proteins and DNA with a method called electron paramagnetic resonance (EPR). On the following pages (shaded in gray) we give the interested reader a short review of this powerful technique.

EPR and Radiation

For the interested reader we give a short presentation of a powerful technique used for studying unstable radiation-induced products known as free radicals. Those not interested can skip these pages.

In EPR (*electron paramagnetic resonance*), an irradiated sample in a strong magnetic field is exposed to microwaves. Under conditions which satisfy the electronic resonance of the free radical, microwaves are absorbed and this absorption yields valuable information about the molecular damage induced by radiation.

The intial products formed by radiation are mainly *"free radicals"*. These products are characterized by having an odd number of electrons and are "paramagnetic"; i.e. they have a magnetic moment μ, which is given by the expression:

$$\mu = g\beta S$$

Here g is a characteristic constant, β is the Bohr magneton (a unit of magnetic momentum) and S is the spin of the electron.

Molecules with magnetic moments behave like small magnets in a magnetic field B. Their energy in the field is given as:

$$\mu \cdot B \text{ or } g\beta S \cdot B$$

S may have one of two values, $\pm 1/2$. Thus, there are two possible energy states. This is the key to understanding the EPR technique. Free radicals in a magnetic field are divided into two groups (the magnetic moments either oppose B or align with B) each group having a different energy.

When the sample containing free radicals is exposed to microwaves of the correct resonance energy, transitions are induced from one energy state to another. The requirement for this is that the microwave energy is exactly equal to the energy difference between the two states. This requirement can be written:

$$h\nu = g\beta B$$

In this equation, the energy is the product of Planck's constant (h) and the microwave frequency (ν). For a magnetic field of 0.33 tesla (3,300 gauss) resonance occurs at a microwave frequency of about 10 GHz. This frequency corresponds to a wavelength of 3 cm (the same wavelength region as a radar system).

An EPR-signal or spectrum is observed when the magnetic field is swept and the resonance conditions are fullfilled. Thus, the EPR spectrum of a sample exhibits the absorption of microwaves versus the magnetic field. The shape of the spectrum yields information on the environment of the unpaired electron and if the electron interacts with neighboring protons and other nuclei. These interactions often make possible the identification of the initial and secondary radiation products. Experiments can be carried out on irradiated samples at a very low temperature where the primary products are frozen and, thereby, stabilized. By subsequent warming the products are released and secondary reactions studied. Thus, the EPR experiments yield :

1) identification of the free radical products formed,
2) the concentration of radicals per unit dose (chemical yield), and
3) the secondary reactions spawned by the initial radicals.

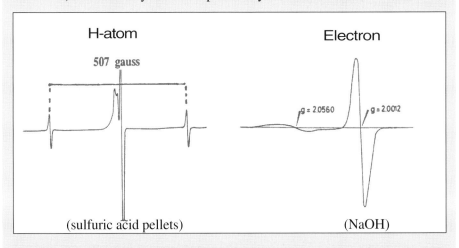

Figure 12.2. Two EPR-spectra which show the "fingerprints" of the two important radiation-induced products from water, namely H-atoms and hydrated electrons. The two lines with a splitting of 507 gauss is the "fingerprint" of the H-atom, and the line with the g-value of 2.0012 is the spectrum of a stabilized electron. The species are trapped in frozen aqueous solutions at the temperature of liquid nitrogen (−196 °C). (Henriksen, 1964)

In order to give an idea of the technique, two spectra are presented in Figure 12.2. The spectra exhibit two of the most prominent radiation products in water (H-atoms and electrons). Frozen pellets of sulfuric acid (to the left) and NaOH (an alkaline solution) have been irradiated and observed at the temperature of liquid nitrogen (–196°C). A hydrogen atom gives a spectrum consisting of two lines with a separation of 507 gauss (or 50.7 millitesla).

An electron, stabilized in water, has an EPR-spectrum consisting of a single line. The position of the line in the magnetic field corresponds to a g-value of 2.0012.   The g-value tells us that the electron is kept in a cage of water molecules. If the electron was completely free the g-value would have been 2.0023.

In the case of radicals formed from proteins, DNA or the constituents of these macromolecules, the spectra are much more complicated. Through advanced techniques such as EPR, valuable information is being learned about the fast chemical processes initiated by ionizing radiation in biomolecules.

Radiation Chemistry

Radiation chemistry involves the irradiation of aqueous solutions and the identification of products and information about the amounts of the products formed (the chemical yields). The formation of polymers via radiation-induced radicals is a part of this research field.

Upon the irradiation of water, radical ions and excited molecules are formed. From these intitial products, secondary free radicals are formed. There are two different pathways :

1. Excitation

$$H_2O + \text{ionizing energy} \Rightarrow H_2O^*$$

H_2O^* represents an excited molecule. The excited molecule is very unstable and loses its extra energy rapidly by a variety of pathways. One pathway is bond cleavage of the excited water molecule, resulting in the formation of H and OH. Both of these free radical products have been identified using the EPR-technique described above.

2. Ionization

$$H_2O + \text{ionizing energy} \Rightarrow H_2O^+ + e^-$$

The primary products H_2O^*, H_2O^+ and e^- give rise to three reactive radicals: OH (hydroxy radical), e^-_{aq} (aqueous or hydrated electron) and the H-atom. In a neutral solution (pH = 7), the relative amounts of these radicals formed are 2.6:2.6:0.6, respectively.

Note that the electron ejected by an ionizing event, e^-, is distinct from the aqueous electron, e^-_{aq}. The e^-_{aq} is an electron that is enveloped by a number of water molecules and is relatively stable (for up to milliseconds). The e^- is a "dry" electron that still retains some of the excess kinetic energy acquired from the ionizing event. The dry electron can be solvated to form e^-_{aq}, or trapped in a frozen matrix (see Figure 12.2), or react with a biomolecule such as DNA (see Figure 12.3).

All the initial water radicals have been observed with the EPR-technique. As early as 1955, R. Livingston, at Oak Ridge, showed that H-atoms were formed in frozen solutions of sulfuric acid, H_2SO_4.

In 1954, R. Platzmann suggested that the aqueous electron is a radiation product. This theory was confirmed by J. Boag and E. Hart in 1962. They observed the absorption spectrum of the aqueous electron in pulse radiolysis experiments. Two years later the hydrated electron was observed using EPR by T. Henriksen, in experiments on frozen solutions of NaOH.

The three primary products formed in water are the starting point for a number of radiation-induced effects in biological systems. One of the goals in radiation research is to follow the processes which take place as these intial products react and yield changes to important biomolecules such as proteins and DNA.

Radiation Damage to Proteins and DNA

A protein consists of amino acids bonded together in long chains called *polypeptide chains*. Altogether, 20 different amino acids can be combined in a large number of ways. Radiation damage and changes to the different amino acids, as well as to shorter or longer peptide chains, have been studied. Many proteins are enzymes, the "workhorses" in the cell operating as catalysts for a number of biochemical processes. The effects of radiation on enzymes have been studied.

Enzymes can be irradiated in solution or in the dry state. Measurements are then made on the degree to which the enzyme activity is lost. The loss of activity is related to chemical alterations induced in the protein structure. In the case of the direct effect, it is known that the damage is often localized to particular regions of any given protein. Since the primary absorption of energy is evenly distributed, this means that radiation-induced damage actually migrates through the protein and "settles out" at specific sites. These are called energetically-favorable trapping sites.

Although protein is damaged by radiation, the consequences are generally **not** significant. This is because proteins exist in multiple copies and, if needed, new copies can be generated using the information stored by the DNA (deoxyribonucleic acid). But what if the DNA is damaged?

DNA

The effects of radiation on chromosomes, genes and DNA has been a major focus of research. The reason for listing these three names together is that genetic information is encoded by DNA molecules that in turn are packaged as chromosomes. Scientists have studied the effects of radiation, chemicals, ultrasound and UV that alter the molecular structure of DNA. Of considerable importance is the correlation between the DNA damage and the subsequent biological effects. A correlation is anticipated because the information carried by DNA is essential for cellular replication and differentiation.

In order to discuss radiation damage to DNA, it is necessary to first give some details about DNA structure.

The research done to understand the structure and significance of the *hereditary molecule* is exciting and a number of Nobel prizes have been awarded to individuals working in this area. In the years 1951 to 1953, work on determining the molecular structure of DNA reached a milestone of epic importance. A biologist, James D. Watson, and a physicist, Francis Crick, working at the Cavendish laboratory in Cambridge, England formulated the double helix model for the three dimensional structure of DNA. Publicaton of their paper "Molecular Structure of Nucleic Acids" in *Nature,* in 1953, brought them instant acclaim. The structure provided immediate insight into the relationship beween DNA's structure and its function. Their DNA model was based on X-ray diffraction studies. Concurrently, experimental x-ray work was carried out in Maurice Wilkins' laboratory in London by Rosalind Franklin. Unfortunately, Rosalind died in 1958, only 37 years old, and she never earned the Nobel prize she deserved.

DNA – The heredity molecule

Watson and Crick presented the DNA model in the English Journal *Nature* on April 25, 1953. The model was relatively simple and provided insight into how genes work and how heriditary information is transmitted.

Nucleic acids were discovered by Friedrich Miescher more than 100 years ago. It was realized early that chromosomes contained genetic information but it was first through the work of Griffith and Avery on bacteria that DNA was found to be the key molecule.

In 1928, Fred Griffith demonstrated that one type of pneumococcus (called R) could inherit the properties of another type (called S) by attaining an extract from dead S-bacteria. In 1944, Oswald Avery, Colin MacLeod and Maclyn McCarty showed that the "effective" substance in the extract was the DNA-molecule. Today this is so well known that it is difficult to appreciate the magnitude of these scientific achievements.

DNA is a long molecule (a polymer) which has 6 different building blocks, the 4 bases; cytosine (C), thymine (T), guanine (G) and adenine (A), as well as a sugar molecule (S) and a phosphate group (P). Phosphate, sugar and a base form a nucleotide and DNA is made up of two long chains of nucleotides.

If all the DNA in a single human cell was tied together and stretched out, it would be approximately 2 meters long. One strand of DNA duplex binds to the other strand through weak bonds (called hydrogen bonds) that extend between base pairs. Thus, C binds specifically with G and A binds with T forming the CG and AT base pairs. The double helix contains about 6 billion basepairs. Three adjacent bases (a triplet or codon) on one strand code for a certain amino acid in a protein. If a protein consists of 200 amino acids, the DNA must code for it using at least 600 bases, i.e., 200 triplet codons. **This is a gene**.

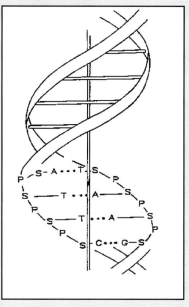

If an error arises in one of the bases, a "wrong" amino acid may be inserted into the protein, changing its properties. An example of an error of this type can cause "sickle cell anemia", a dreaded sickness.

Human beings vary because of differences in the DNA base sequence. This can be demonstrated in so-called DNA-tests. Such tests are now used for positive identification of people (for example in criminal court cases).

Scientists around the world are engaged in research on how external agents such as radiation and chemicals induce changes in DNA. Some changes kill the cell, others change the cell and cause cancer, while other changes are without observable effects.

Scientists involved in the "hereditary molecule"

*Oswald
T. Avery*

The Rockefeller
University Archives

*Maclyn McCarty (l.)
and
Colin M. MacLeod*

The Rockefeller
University Archives

A milestone in molecular biology was passed when Avery and his two coworkers MacLeod and McCarty in 1944 demonstrated that DNA was the hereditary molecule. They worked with two types of pneumococcus. Already in 1928 F. Griffith had shown that it was possible to transform one type of bacteria into the other by an extract of "dead" materials from the first one. In the work *Studies on chemical nature of substance inducing transformation of pneumococcal types* Avery *et al.* showed that DNA was the key substance.

Sven Furberg determined the structure of the nucleotide cytidine. He found that the baseplane was perpendicular to the sugar molecule. Based on this, in 1952, he proposed a single stranded DNA-model.

Rosalind Franklin carried out valuable X-ray diffraction studies in Wilkins' laboratory in London on DNA. She observed two types, A and B, each having a helical structure.

Few, if any papers have had such impact as the single page Watson and Crick published (*Molecular Structure of Nucleic Acids*) in *Nature* in 1953. Watson was only 25 years old and Crick was 37.

Sven Furberg

Rosalind Franklin

F.H.C. Crick

J. D. Watson

Courtesy of CSHL Archives

© The Nobel Foundation

© The Nobel Foundation

It is interesting to note that the Norwegian physical chemist Sven Furberg worked on the structure of DNA components. In the work on cytidine he found that the base plane (the plane of cytosine) was perpendicular to the sugar molecule. Based on this, he suggested a model of DNA consisting of a single stranded helix. The model consisted of a long chain of sugar molecules (marked S) and phosphate groups (marked P); –S–P–S–P–S–P–. The base planes were perpendicular to the axis of the helix and the distance between the base planes was 3.4 A (1 Angstrom = 10^{-10} meter). This model was correct in most aspects but lacked the important idea of a two stranded helix.

The basis for the genetic code is given in the model of Watson and Crick as a double stranded helix. The genetic information is simply the order of the bases in the helix (the primary structure). A virus typically requires a few thousand steps (bases) in the DNA-ladder. If this DNA molecule is stretched out it extends a couple of micrometers. In order to write down the information contained on the viral DNA, using one letter for each base (or step in the ladder), a couple of book pages would be needed.

In order to build a bacterium, more information would be needed, approximately 50 pages. A human being however, is far more complicated and, in order to write down the order of bases, a book of about one million pages would be needed. The research world is now engaged in "writing this book" in a large study called *the human genome project*. Chromosome 22 has recently been sequenced and, within the next few years, the genetic code for human DNA will be known.

In a human cell, the DNA-thread is packed into 46 units (the chromosomes). With the use of particular methods it is possible to study the chromosomes under the microscope when the cells are in division (in mitosis).

Growth takes place by division of cells. Each cell goes through a cycle (see page166) and, before division, the content of DNA must double. The goal is to make the new DNA identical to the old DNA in this process called *replication*. It is in fact, fantastic to realize that very few significant errors arise when a book of a million pages is printed over and over again. Some errors actually arise routinely but they are immediately recognized and repaired. This editing and repair is done by a number of molecules called repair enzymes.

If an error arises that is not repaired or if it is misrepaired and the cell still divides, this is a mutation. If the mutation occurs in an ordinary cell in the body it is called a *somatic mutation*. If the cell is capable of reproduction, it may lead to cancer.

A mutation in a sex cell is called a *genetic mutation*. Such mutations can take place spontaneously and people have for a long time speculated about the mechanisms. One way of producing genetic mutations is by radiation.

The evolution of species requires mutations. Thus, a slow development of the species is based on accidental mutations. Research work seems to indicate that the frequency of the spontaneous mutations is increased by radiation and it has been a long range goal to determine the dose which doubles the mutation rate for animals and humans. This is called the *doubling dose*.

Radiation damage to DNA

The most important types of damage to the DNA molecule, induced by radiation, are presented in Figure 12.3.

There are four common types of radiation damage to the DNA molecule:

1. Single strand breaks

A single strand break is simply a break in one of the sugar-phosphate chains. This damage is usually simple to repair and, in experiments, it has been shown that approximately 90% of the single strand breaks are repaired in the course of one hour at 37°C.

2. Double strand breaks

This type of damage involves both strands of the DNA helix, which are broken opposite to each other or within a distance of a few base pairs. If you look at Figure 12.3, you may imagine that this damage would kill the cell and in experiments with bacteria a correlation is found between double strand breaks and cell death.

The way that chromosomes have been described it would seem impossible for a double strand break to be repaired. However, they actually are. The DNA-molecule is packed together with proteins supporting the structure and preventing the pieces from falling apart, even when breaks occur on both strands of the helix. There are in fact a number of mechanisms that complex organisms (such as humans) have evolved for repairing double strand breaks. But as one might guess, this type of break is more difficult to repair and does correlate with observable damage to chromosomes.

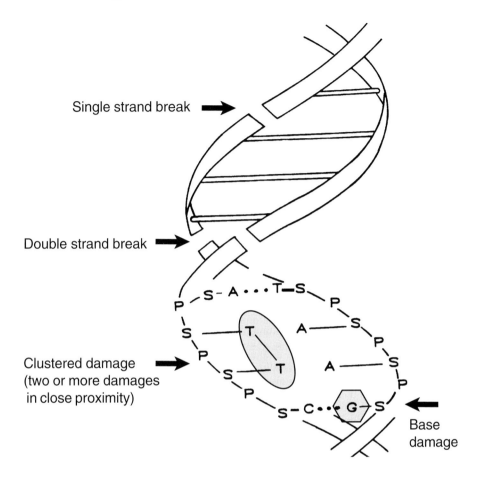

Single strand break

Double strand break

Clustered damage
(two or more damages
in close proximity)

Base
damage

Figure 12.3. The DNA molecule with 4 common types of radiation damage (see text).

3. Base damage

Experiments indicate that radiation sensitivity varies from one base to another. After an initial ionization, rapid electronic reorganizations take place with the result that the damage is transported to certain regions of the macromolecule. The base guanine is particularly sensitive.

Damage to a base is one of the starting points for a mutation. If a base is changed, information may be lost or changed. As the result of a missrepair or no repair, the altered triplet codon (see p 159) is likely to lead to insertion of the incorrect amino acid in the protein. In turn, the changed protein might not function properly. It is known that certain illnesses have their origin in base changes and one hypothesis is that some radiation-induced cancers may be due to base changes. This is one reason why it is important to get information about the processes that take place in the cell at the molecular level.

4. Clustered damage

Clustered damage is what makes ionizing radiation quite different from other agents that cause DNA damage. One example of clustered damage is shown in Figure 12.3.   In this example, two adjacent bases, T and T, on the same strand have been chemically altered.   This is but one out of a myriad of possibilities. All the possibilities have the common feature that two or more damaged sites lie in close proximity to one another.

There is a second example of clustered damage shown in Figure 12.3, the double strand break which was just discussed above. Double strand breaks result whenever two or more single strand breaks occur in a cluster with at least one break occuring on each of the opposed strands. Double strand breaks are an important class of clustered damaged. Their importance stems from the fact that they are readily observed, making it possible to learn a lot about them, and from the observation that they are difficult to repair correctly.

The difficulty posed by clustered damage is that one of the strands is needed to replicate the adjacent strand. When both strands are damaged at the same site there is no template to work from. This is in sharp contrast to damage such as a single base alteration or a single strand break. These single damages are readily repaired by using the information on the opposite strand. Indeed in the absence of radiation, or any other outside agent, your DNA is being continually damaged and continually repaired. The fidelity of that repair is extremely high because endogenous damage is not clustered but instead occurs in separated sites of single damage. Single damage repair is routine but clustered damage repair is a major biological challenge and is prone to error.

Look back at Figure 4.1 on page 33. Note that, by its very nature, ionizing radiation deposits energy in tracks that are composed of clusters of ionizations. It is this quality that makes the biological effects of radiation unique.

Radiation Damage to Cells

It is possible to grow human cells in a tissue culture laboratory. These cells can be irradiated and treated in different ways without any consequences to the donor. A tissue culture laboratory is important in cancer research and radiation research.

In radiation experiments the goal is to determine dose-effect curves for different biological end effects (for example, cell death). Since it is difficult to observe dead cells the dose effect curves are given as *survival curves*. That is the number of cells that survive a certain dose.

The cells usually grow asynchronously, that is the cells are in different phases of the cell cycle. The radiation sensitivity of the cells changes from one phase to another (see next page).

Experiments with beams of particles which can be directed to different parts of a cell demonstrate that the nucleus is more radiation sensitive than the cytoplasm. When asynchronous cell cultures are irradiated with a dose of 5 Gy a decrease in mitosis is observed (mitosis is the division of the cells, and consists of the phases: prophase, methaphase, anaphase and teleophase). The fraction of cells which are in mitosis ("the mitotic index") will decrease. This means that the growth in the number of cells has slowed. After a few hours the mitotic index will again increase, reach a maximum, and then decrease.

Cell death induced by radiation can be divided into two groups:

1. If a cell dies after the first mitosis it is called *mitotic death* or *reproductive death*.

2. If the cell dies before reaching the first mitosis it is called *interphase death.*

Cells which survive large doses very often have chromosome abnormalities.

The cell cycle

A living cell goes through several stages from the moment it is "born" by cell division until it divides, forming two new cells. Before division takes place the cell content of DNA must double. This cell cycle can be divided into 4 stages (see figure) and the radiation sensitivity differs for each stage.

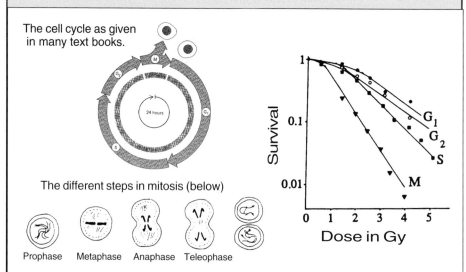

The cell cycle as given in many text books.

The different steps in mitosis (below)

Prophase   Metaphase   Anaphase   Teleophase

In the **M-phase** (mitosis) the cell divides and a new cell starts its cycle. The mitosis itself is divided into different phases such as; prophase, metaphase, anaphase and teleo-phase. The M-phase in mammalian cells takes 1 to 2 hours.

In **S-phase** the content of DNA is doubled.

Between M- and S-phase we have two phases or gaps which have the names G_1 and G_2. The G-phases and the S-phase combined is called the "interphase". The time spent in the different phases is approximately as shown in the figure.
Methods are available to study the processes controlling the cell cycle. The sensitivity to radiation varies according to the phase. One

example of this is shown in the above figure In order to do experiments of this type, researchers work with cells growing synchronously (all cells are in the same phase). Cells in culture grow asynchronously, where cells are in all phases in a ratio corresponding to the time interval for the phase. The fraction of cells in mitosis is called the "mitotic index". This parameter is studied in radiation experiments.
The figure above is for HeLa-cells (a human cell used in research). Other cell types exhibit different survival curves. In general, the cells are most radiation sensitive in mitosis.
Since growing tissues have different ratios of cells in each phase, the radiation sensitivity varies from one type of tissue to another. This is important in radiation therapy.

Protection Against Radiation

During the 1950s and 1960s many scientists were engaged in attempts to find compounds that would provide some protection against radiation. It is not difficult to see the advantages of "a pill" which protects against radiation. The idea was to administer compounds that could change the cells to reduce their radiation sensitivity.

A number of organic compounds do provide some protection against radiation if given *before* the exposure takes place. Most of these compounds contain sulfur, either a thiol (a -SH compound) or a disulfide (a -SS- compound). In animal experiments these compounds may yield protection against radiation by a factor of 2 to 3. This means the dose must be increased by a factor of 2 or 3 to provide the same effect.

There is on-going research to understand the chemistry of radiation protection compounds and develop these compounds on the molecular level so that they can be placed within our own cells to safeguard them from radiation damage.

Oxygen Effect and Sensitizing Compounds

Experiments are also being carried out to find compounds that can increase the radiation sensitivity of malignant cells so that they are destroyed with lower radiation doses.

When radiation is used in the treatment of cancer the purpose is to kill the cancer cells. At the same time the healthy cells should survive. One of the problems encountered is the *oxygen effect*. In the center of a tumor the cells are far away from the bloodstream and are deprived of oxygen. They are more or less hypoxic.

The oxygen effect is an experimental observation that cells with low oxygen content are more resistant to radiation than cells with a normal oxygen content. So, for treatment of a tumor, a radiation dose that is sufficient to kill the "normal" cancer cells may be too small to kill the hypoxic cells in the middle of a tumor (see Figure 12.4).

A tumor model

Blood vessel

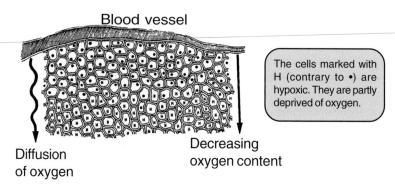

The cells marked with
H (contrary to •) are
hypoxic. They are partly
deprived of oxygen.

Diffusion
of oxygen

Decreasing
oxygen content

Survival curves

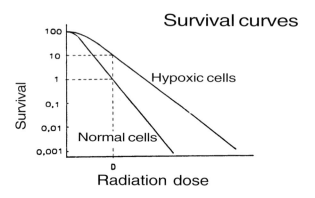

Figure 12.4. Above is a model of a tumor. The cells are supplied with oxygen diffusing from the blood vessel. The oxygen content in the cells decreases with the distance from the blood vessel. The graph shows the survival curves for cells with normal oxygen content as well as for hypoxic cells.

Cells in the middle of a tumor may be at some distance from the blood vessels and the oxygen supply will decrease. In some cancers, tumors with a diameter of as small as 2 mm contain regions with hypoxic cells. These cells can survive the radiation treatment and subsequently develop into a tumor.

Several methods have been used to increase the radiosensitivity of hypoxic cells in a radiation treatment:

1. The radiation treatment is protracted. The total dose is split into a series of smaller doses separated by intervals of rest; for example, one dose per day for two weeks. This procedure is used because tumor cells with a normal oxygen content will die and disappear in the first part of the treatment. This will yield opportunities for the deeper lying hypoxic cells to regain a normal oxygen content, becoming more sensitive to radiation in the latter part of the treatment.

2. Attempts have been made to irradiate the tumor with particles (protons, α-particles, heavy ions) having a high LET (linear energy transfer). The oxygen effect decreases with increasing LET.

3. Attempts have been made to increase the oxygen content during treatment with the help of oxygen pressure chambers and gas masks where the patient breathes oxygen-rich air. This method has not been a success.

4. In more recent years attempts have been made with chemical compounds that can be added to the hypoxic cells. The point is to increase the radiation sensitivity. It appears that nitroxide compounds do have a sensitizing effect. A good example of one of these compounds is "misonidazole".

There were great hopes for the use of sensitizing compounds in the treatment of cancer. However, at the present time there is not a sufficient net positive effect in using such compounds.

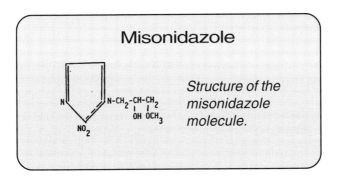

Misonidazole

Structure of the misonidazole molecule.

Radiation and Chromosomal Aberrations

In recent years, methods have been developed that make it possible to study the occurance of radiation damage to DNA. This is done by observing alterations in the shape of chromosome. Shapes that deviate from normal are called chromosomal aberrations.

Circulating lymphocytes are used in detecting chromosomal aberrations. These cells are distributed throughout the body and are all in the same phase of the cell cycle, the resting phase. In the procedure, lymphocytes taken from the blood, isolated from the whole blood, and stimulated to grow, i.e. they are triggered to enter into the normal cell cycle.

The cells are cultured for 48 hours at 37°C and are harvested when they are in the part of mitosis called metaphase. Then the chromosomes are prepared using standard methods and can be studied under a microscope to observe any changes.

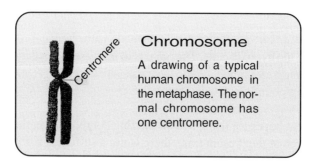

Chromosome

A drawing of a typical human chromosome in the metaphase. The normal chromosome has one centromere.

Several types of aberrations are observed. The changes induced by radiation are "ring formation" (the chromosome seems to form a ring) and "dicentric chromosomes" (the chromosome has two centromeres instead of one).

The mechanisms for the formation of aberrations are not yet fully understood. To some degree, they involve strand breaks in the DNA-molecule. While the biological effects connected with chromosome changes are also not fully understood, there is a correlation between cancer induction and the incidence of aberations. In addition, the formation of chromosomal changes can be used to attain information on the radiation dose. For example, the Japanese fishermen on the Fukuru Maru who were irradiated in 1954 had an increased number of chromosomal aberations in their lymphocytes when examined in 1982, twenty eight years after the exposure. Similar chromosomal aberrations have been noted in the A-bomb survivors and in accident victims. This "biological dosimeter" (chromosomal aberations) can be used in a dose range from about 0.2 Gy to 12 Gy.

Attempts have been made to find chromosomal aberrations in groups of people exposed in the Chernobyl accident and after the atmospheric nuclear tests in the 1950s and 1960s. The doses in this case would be a few mGy distributed over several years; no significant increase in aberrations has been observed.

Repair Processes

In 1960 M.M. Elkind and H. Sutton at the National Institutes of Health published a paper which demonstrated cellular repair processes. When they irradiated Chinese Hamster cells with a dose of 10 Gy the survival was 2 per 1,000. If, however, they split this dose into two equal doses of 5 Gy with an interval of 2 hours between exposures, the survival was 8 per 1,000. If the time lapse between the two doses was shorter the survival was reduced. Elkind and Sutton interpreted the results in the following way; the first dose of 5 Gy killed a number of cells whereas other cells attained damage that they called "sublethal damage". In the time interval between the two doses the damage could be repaired and the cells were "healthier" when the next dose hit. This can be compared to a boxing match where one of the boxers sustains a number of hits but is saved by the bell. Between rounds the crew works in order to get the boxer fit to continue the fight.

Cells have repair systems. This is a necessity for survival. The crew working on repair in our cells are enzymes. It is the job of some enzymes to detect DNA damage while others are called upon to repair the damage. The repair processes can be divided into three types:

• The specific site of damage is repaired. In this case the enzymes work right at the damaged site. The original base sequence is preserved.

• The whole stretch of DNA containing a damaged site (or sites) is removed and replaced, preserving the native sequence.

• The damage is ignored during replication; it is by-passed. With luck the correct base will be inserted or, if incorrect, it won't matter. Because this type of repair is error prone, it is held in reserve in case the higher fidelity repair systems miss, or cannot cope with, the damage. For this reason, it is aptly called "SOS" repair.

In Figure 12.3 some of the main types of damage to the DNA-molecule are shown. One important repair mechanism is "excision repair". This repair mechanism involves enzymes that cut out the damaged part of DNA and replaces it with a new undamaged part. One of the scientists working on understanding repair processes is Gunnar Ahnstrøm at the University of Stockholm. His imaginative drawing showing the essential elements of DNA repair is shown in Figure 12.5.

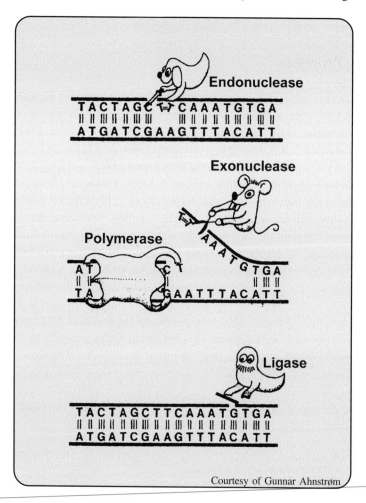

Courtesy of Gunnar Ahnstrøm

Figure 12.5. Repair of pyrimidine dimers in DNA. The process includes several steps and altogether four enzyme groups. If you look into the details you will find that Prof. Ahnstrøm has included an error in base pairing. Can you find it?

The repair mechanism in Figure 12.5 includes the following steps:

1. Recognition. It is important to have enzymes that can recognize the damage and signal for help.

2. Cutting of the DNA-strand. It is a requirement that specific enzymes, like the endonucleases, can cut the DNA-strand in the neighborhood of the damage.

3. The damaged part is removed and rebuilt. Exonuclease and polymerase are key enzymes. The former cuts out the damaged part and the later replaces it with a new undamaged part.

4. Joining. The repair is finished when the ligase enzyme joins the cut DNA-strand back together.

The repair system, outlined above, is found in humans and microorganisms. An array of repair mechanisms are used to repair not only radiation-induced damage but also damage stemming from a multitude of other agents, including the routine damage that occurs as part of normal cell function. A repaired cell divides and functions in a normal fashion.

What happens when the repair system fails or is too weak?

When we are out in the sun and exposed to UV light, adjacent pyrimidine bases (C or T) become fused together (pyrimidine dimers) in the skin cells. Normally, our repair mechanisms are intact and can repair the extra damages from the UV-radiation. However, there is a genetic ailment for which the above described repair system is too weak to repair all of the extra damage caused by the sun. The ailment is called *Xeroderma Pigmentosum*. It has been found that the enzyme endonuclease is weak and often fails to do its job. The result is that people with this genetic defect develop skin cancers that are often lethal.

During the past few years some interesting research has been published dealing with the process called *adaptive response*. There is some evidence that small doses of radiation (up to 100 mGy) can increase cellular resistance to genotoxic effects. The mechanisms by which this could occur are either stimulation of the repair system or cell death by apoptosis. In the next section, examples of such experiments will be described briefly. If it is true that small doses of radiation initiate both positive and negative effects in cells, the implications for health (i.e., the dose–effect curve discussed in Chapter 11) are very important.

Adaptive Response

The repair system in cells is absolutely necessary for survival. Research demonstrates that cellular resistance can be increased by small stimulating doses. The radiation community has named it *adaptive response.*

Experiments exhibiting adaptive response started in 1984 with the work on human lymphocytes by G. Olivieri, J. Bodycote and S. Wolff at University of California. The lymphocytes were cultured with ^{3}H-labeled thymidine that was incorporated directly in the DNA-molecule and served as a source of low-level chronic radiation. The cells were then irradiated with x-rays to a dose of 1.5 Gy and the yield of chromosome aberrations recorded. It was found that the number of chromosome aberrations was *fewer* after exposure to both sources (tritium β-particles as well as x-rays) than after x-rays alone. These results showed that low levels of radiation can trigger or induce increased repair of radiation induced chromosome breaks.

Throughout the 1990s a large number of experiments have been published on different systems that demonstrate an adaptive response. A number of end points have been studied such as cell killing, micronucleus formation, induction of chromosome aberrations, induction of mutations and neoplastic transformations. The adaptive response has been detected when cells have been exposed to a small dose (10–100 mGy) and then challenged with a much higher dose. Let us briefly mention a few points:

- The amount of chromosome damage in lymphocytes can be reduced by up to 50% if a dose of 10 to 150 mGy is given to the cells before a larger dose of 1.5 Gy is given.

- In order to attain a maximal effect, the stimulating dose must be given 4 to 6 hours ahead of the large dose. The effect seems to last through a couple of cell cycles.

- Adaptive response has also been found in the case of mutations in the fruit fly (drosophila melanogaster). The number of mutations was reduced when a small dose of 200 mGy was given before a large dose of 4 Gy.

There is evidence that the adaptive response is mediated through the induction of repair enzymes. For example, the response is inhibited by the prevention of protein synthesis as well as by inhibition of polymerase that rejoins DNA-breaks. The research also indicates that the immune system in mice can be stimulated by small doses. Furthermore, small doses of radiation seem to trigger the formation of proteins that can stimulate the growth of cells in the thymus and spleen. This, in turn, would have a positive effect on the formation of antibodies.

J.L. Redpath and R. J. Antoniono reported in 1998 on experiments that demonstrate an adaptive response against spontaneous neoplastic transformation in cell cultures (doses down to 10 mGy). These exciting experiments where no large dose was used, provide evidence that radiation in itself may have positive effects.

Conclusion

Radiation in small doses may stimulate a number of different processes, not all of which are deleterious. Radiation stimulates both the repair system and apoptosis, the system that removes malfunctioning cells.

If one accepts the conclusion that radiation in small doses intiates both negative and positive processes, an important question is raised. Which is dominant? If the combined effect is positive, the outcome would be called *hormesis*.

Whether or not the final outcome is positive (hormetic), the adaptive effect raises an important issue. How do we make risk-benefit decisions in the low dose regime (in the range up to 100 mGy)? For example, should the adaptive effect be taken into consideration when deciding how to respond to the marginal increases above background produced by the Chernobyl accident?

Because experimental results raise the possibility that small doses of radiation are beneficial to human health, the linear no-threshold model for risk assessment (presented in Chapter 11), needs to be re-examined. This is in fact what is currently being done by both research scientists and policy makers. It seems logical that new exposure limits would be based on the average natural background levels and their variation (2 to 10 mSv per year).

Genetic Damage

In 1927, Herman Muller observed that ionizing radiation causes mutations. He worked with the fruit fly (drosophila) and used x-rays and found an increased mutation frequency in the X-chromosome. This was the starting point for radiation genetics (see next page).

Radiation may damage the DNA-molecule. It was mentioned above that if the damage occurs in one of the cells of the body, and the damage is not repaired, a somatic mutation may result. On the other hand, if damage occurs in an ovary or sperm producing cell, then a genetic mutation may result. A genetic mutation is likely to give observable changes in subsequent generations.

Most of the knowledge about radiation-induced mutations was obtained through laboratory experiments. For example, irradiated mice have been studied. At Oak Ridge one experiment used about 7 million mice. "The megamouse experiment" provided evidence for seven different types of mutations (changes in the fur color, the ears, etc.). Radiation had increased the number of mutations. The results can be summed up as follows:

- The radiation sensitivity for the different types of mutations varied by a factor 20.

- With mice, a significant dose-rate effect was found, the mutation frequency increasing with increasing dose rate. Consequently, a protracted dose yielded a smaller effect. This result was not found in the fruit fly experiments.

- Male mice were more radiosensitive.

- The number of mutations for a certain dose decreased with the lapse of time between radiation and conception. This seems to be equal for the two sexes. For humans, the lesson might be that if the gonads receive a sufficiently large dose a planned conception should be postponed a few months.

- Spontaneous mutations always take place. One possible source of these mutations is background radiation. The animal experiments seem to indicate that background radiation accounts for about 2% of the spontaneous mutations. A radiation dose of about 1.5 Gy increases the mutation frequency by a factor of 2 (*the doubling dose*).

Fruit flies
Radiation – Genetics

The fruit fly with the scientific name *Drosophila melanogaster* has played a significant role in genetic studies. The reason is that the fruit fly is an excellent model system; it is inexpensive, easy to work with and exhibits genetic changes that are easy to observe.

Herman J. Muller used the fruit fly in 1927 when he reported on "true gene mutations". In the work *Artificial Transmutation of the Gene* published in *Science,* Muller described how x-ray irradiation yields observable mutations in the X-chromosome. He earned the Nobel price (in 1946) for this work.

It was mentioned that the genetic changes are easy to observe and the changes in question are, for example, the color of eyes and defects to the wings. One easily observable mutation is given in the figure below. If you compare this fly with the normal one, you can easily observe curled wings.

The fruit flies are rather small and the observations are made using a magnifying glass. The flies can be kept in glass jars, and may be anesthetized with ether in order to keep them quiet during observation.

Normal fruit fly

Courtesy Radiation Research Society

H.J. Muller

A mutation

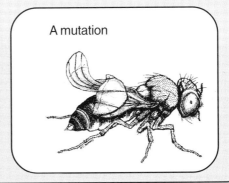

The mouse experiments yielded results that were somewhat different from the experiments with fruit flies and have moderated the concern about the genetic effects of radiation. Two reasons for this are: first, the doubling dose seems to be far larger than that previously assumed, and second, the mutation frequency depends on the dose rate.

In the case of humans, the data on atomic bomb survivors in Japan is informative. Approximately 70,000 children have been registered where the parents received radiation doses of about 350 mSv. This cohort has been studied carefully for the incidence of stillborn, child death, deformities, death before the age of 26, an abnormal number of chromosomes, and changed ratio of girl/boy. In a book containing data up to 1991, J. V. Neel and W. J. Schull concluded that the doubling dose for humans is about 2 Sv for acute radiation and about 4 Sv for protracted radiation. In the view of these results, genetic effects from the Chernobyl accident are not expected to be detected.

Radiation During Pregnancy

Two different causes may yield deleterious effects to a fetus. In the first place, genetic or hereditary damage is possible (i.e. changes in the DNA molecules to one of the parents). And second, an exposure may occur during pregnancy. The following discussion considers the latter.

Some years ago Alice Stewart and coworkers suggested that the fetus is very sensitive to radiation and that one of the reasons for childhood cancer could be due to exposure during pregnancy. The support for this statement was the observation that 15% of the children that died from cancer before the age of 10 in England had mothers that were x-rayed during pregnancy.

The above hypothesis has been debated. One argument is that the group studied was predisposed to cancer. The fact that the mothers had to go through an x-ray examination during pregnancy was an indication that they already had problems. Further evidence for the theory that the group was at high risk came from two studies. In the first study, the incidence of cancer was also found to be above normal for the sisters and brothers of the exposed group. Second, no increase in the incidence of cancer was found for about 2,000 Japanese irradiated in the embryological stage when the bombs fell on Hiroshima and Nagasaki.

The pregnancy period can be divided into different stages (see Figure 12.6):

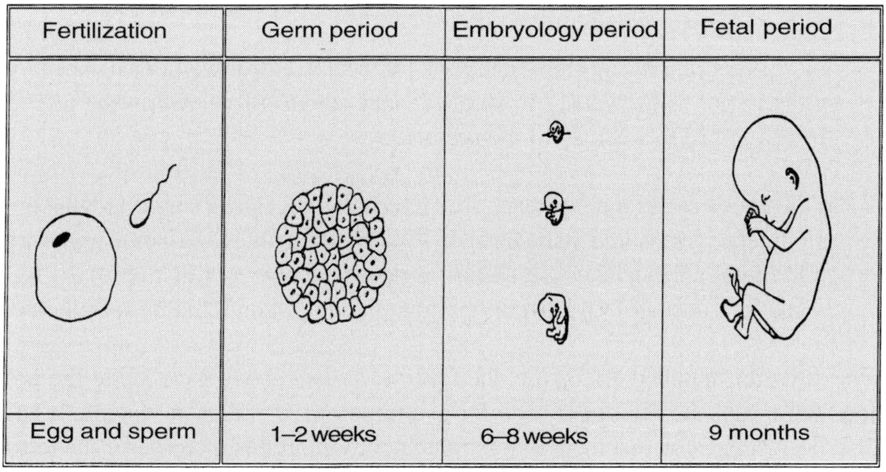

Fertilization	Germ period	Embryology period	Fetal period
Egg and sperm	1–2 weeks	6–8 weeks	9 months

Figure 12.6. Three different stages during pregnancy.

- **The germination period** includes the first days after fertilization. The cells divide rapidly without major differentiation. In a number of experiments with animals, it has been shown that radiation at this stage may result in cell death. The sensitivity decreases with the number of cell divisions. There is no evidence for damage that is carried on by the fetus.

- **The embryological period** goes from the second week to about week 8 or 9. In this period the different organs are formed. External influences yield damage and deformities at this stage. In experiments with animals, deformities have been observed after large radiation doses. No such effects have been observed for humans.

  On the other hand, some chemicals have produced tragic effects. A number of years ago a number of pregnant women used the chemical thalidomide for the nausea that often occurs early in pregnancy. This resulted in deformities.

- **The fetal period** goes from about week 8 to the end of pregnancy. Radiation effects in this stage are mainly connected to the central nervous system.

About 1,600 children were irradiated in the fetal period in Hiroshima and Nagasaki (acute doses of the order 100 to 500 mGy). Of these, 30 were classified as being mentally retarded.

These observations suggest that the risk for mental disorders is about 0.4 per Gy in the period from week 8 to week 15 and about 0.1 per Gy in the period from week 16 to week 25 of the pregnancy.

If the above mentioned data are used uncritically it can be suggested that the intelligence (measured in the form of IQ) is reduced due to radiation exposure in the fetal period. The risk factor proposed involve a reduction of 30 IQ-points for a dose of 1 Gy! In our opinion this suggestion is highly speculative.

The radiation-induced effects on the fetus described above have led to the policy that x-rays are no longer used for examinations of pregnant women. It has also been suggested that the risk for spontaneous abortions increases if the fetus is exposed to doses above 100 mGy in the first 6 weeks.

Chapter 13

Radiation and the Environment

Radiation, both natural background and man-made, forms a part of our environment which must be considered as both a benefit and a risk. In order to judge radiation a number of things must be taken into account, such as:

1. The radiation dose

When a situation occurs with radioactive pollution, it is important to get accurate information about the radiation doses that are involved. It would be a waste of resources to immediately implement countermeasures if the doses involved are smaller than the variations in natural background radiation.

To date no general agreements have been reached with regard to the dose levels where actions should be taken. For example, in the case of radon, the recommendations from WHO (World Health Organization) are that 800 Bq/m³ should be considered as an action level. The equivalent dose involved would be approximately 20 mSv per year.

2. Type of radiation

In order to perform dose calculations and estimate health risks, it is important to have information on the radiation source (e.g., the isotopes involved). As pointed out in previous chapters, there is a clear distinction between γ-emitting isotopes and those which emit α- and β-particles.

3. The amount of pollution

The amount of radioactive material released to the environment should be given in Bq (or in Ci) and not in volume or weight since a large release in weight may contain small amounts of radioactivity and vice versa.

In the example of the Chernobyl accident a large fallout of Cs-137 resulted, both to areas around the reactor as well as far away. In Scandinavia, areas were found with fallout of 100 kBq/m^2. However, the total release of Cs-137 was in fact only a few kilograms (see exercises in Chapter 14). Here, a small amount of material produced an easily measurable amount of radioactivity at a great distance from the accident.

4.   The form and dispersion of the release

It is important to have information on the form of the release; whether it is in the solid state, liquid state, or gaseous state. It is also important to know the dispersion or scatter characteristics in some detail in order to localize so-called "hot" areas. Consider two examples which demonstrate the differences with regard to transport and dispersion of radioactivity:

- **The Chernobyl accident**

This accident caused a large release of radioactivity that reached a height of several thousand meters and was then transported by the wind systems. The fallout was mainly determined by precipitation in the areas where the radioactive "cloud" passed by. After fallout, the different radioactive isotopes still have possibilities for further dispersion via the water systems and plant uptake. This means that the isotopes reached the food chain and gave a large segment of the population an extra dose (see Chapter 9).

- **The submarine Komsomolets**

In 1989, the Russian submarine Komsomolets was sunk in the North Atlantic to a depth of 1,680 meters. It contained a reactor, fission products, and nuclear war heads. The dispersion of radioactivity from this accident is so far not measurable and extra doses to the public are almost nonexistent. Because plutonium is not dissolved easily in water, even as it is released in the future, its dispersion will be very small.

Use of Radiation in Society

Human activities include the use of radiation. Some of these activities may involve a high probability for pollution. If the risk is high, the activity should be stopped. Again, the benefits must be compared with the risks.

Most people would agree that radioactive isotopes used in medical diagnoses and research involve minor pollution problems. However, the large radioactive sources used in therapy and in other activities have resulted in some radiation accidents with fatal outcomes. It is, therefore, necessary to train workers and to implement proper safety routines. This is particularly true within the nuclear power industry where the use of uranium involves risks for the release of radioactive isotopes from the mining of the ore to the final disposal of the waste. Under normal operating conditions, only negligible amounts of radioactivity are released from a power reactor. The public has concerns with waste disposal and the relatively low risk of accidents.

Nuclear Power

In a power reactor, fission energy is used to produce electricity. In the late 1990s, there were approximately 430 reactors in use in 26 countries. World-wide, nuclear power is responsible for 17% of electricity production. And, in 12 countries, it represents more than 25% of electricity production. In France, Belgium and Sweden nuclear power is responsible for more than 50% of electricity production. Nuclear power contributes, at the turn of the century, only 5% of the total energy production in the world. However, the World Energy Council has made several prognoses about the energy need in the next century, based on three different economic growth rates. All scenarios involve a considerable increase in energy production and the use of nuclear power in particular. Furthermore, according to international agreements in Kyoto, in order to reduce the release of greenhouse gases (carbon dioxide – CO_2), the use of nuclear power may be even more important.

A number of different reactor types have been developed. There are thermal reactors (the fission process requires thermal or slowed down neutrons), and fast reactors (the released neutrons are neither moderated nor thermalized).

Energy from the atom

Fission – Fusion

Nuclear power is based on the two physical processes called fission and fusion. Today only the fission process is used, but hopefully fusion will be used in the future. The figure below gives the binding energy in the atomic nucleus. The figure can be used to explain the principles behind the use of nuclear energy.

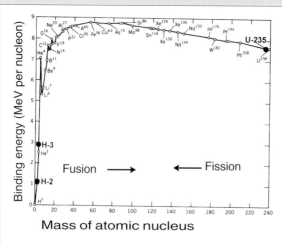

Energy is gained if we can transform a nucleus with a small binding energy into a nucleus with a large binding energy. The figure shows two possibilities.

Protons and neutrons are kept together by strong forces in the atomic nucleus. Along the vertical axis above is given the binding energy per nucleon (mass unit). The figure says that the binding energy increases from about 1.1 MeV (for deuterium) to about 8.8 MeV (for iron). It decreases again toward heavier atoms.

It is possible to gain energy by transforming a nucleus with a small binding energy to another with a larger binding energy. Thus the gain is about 24 MeV when two deuterium atoms combine to form helium (He-4). Upon a fission of uranium into two almost equal parts the gain is approximately 230 MeV.

In conclusion: There are two possibilities for energy production from the nucleus:

1. Fission. A large atomic nucleus is split into smaller units. From the curve above it requires that a heavy nucleus (to the right) be fissionable. Atoms that are fissile are the isotopes U-233, U-235 and Pu 239.

2. Fusion. Energy is gained in a fusion process. The requirement is that light atoms with a low binding energy are used. Hydrogen, deuterium and helium are all atoms that can be used for fusion. The fusion process requires a very high temperature (several million degrees) and is the main energy source for the sun. So far, a controlled fusion process with gain of energy has not been achieved.

The power reactors used today are based on thermal neutrons. They are divided into two groups, namely gas-cooled/graphite-moderated reactors and water-cooled/water-moderated reactors (light water reactors). In addition, there are reactor types between these categories; for example, the Russian water-cooled/ graphite-moderated reactors (the so-called Chernobyl type). There are also heavy water-moderated reactors.

Light water reactors

There are two types of light water reactors: Boiling Water Reactors (BWR) and Pressurized Water Reactors (PWR). In both these reactor types, the uranium fuel and the control elements are placed in a pressurized water container. The energy from the fission is transferred to the water which in turn transmits the heat energy to a secondary cooling circuit.

In a BWR the water is boiling in the container, whereas a PWR operates at a higher pressure (about 150 atmospheres pressure) in order to prevent boiling.

The efficiency for these reactors is approximately 30%. That means about 2/3 of the energy is lost, mainly due to transfer of heat to the cooling system.

Fast breeder reactors

The most abundant uranium isotope in nature is U-238. Only 0.7 % is U-235, which is the uranium isotope used for energy production.

Most of the heavy atoms having an uneven number of neutrons are fissionable and are also radioactive. Small amounts of U-235 are still present, whereas other natural fissionable isotopes have almost decayed away. It is possible, however, to produce other fissionable isotopes, such as Pu-239 and U-233. They are produced in a reactor when neutrons are captured by U-238 and Th-232. If the reactor produces more fissionable materials than are used, it is called a breeder reactor.

In a Fast Breeder Reactor (FBR), the core is enveloped by U-238 or Th-232. Neutrons that escape the core are absorbed by this envelope and fissionable materials are produced. In order to have such a process running the core, the enrichment of U-235 must be 15 to 20%.

The fissionable plutonium is formed according to the following reactions:

$$^{238}_{92}U + ^{1}_{0}n \Rightarrow ^{239}_{92}U + \beta \Rightarrow ^{239}_{93}Np + \beta \Rightarrow ^{239}_{94}Pu$$

The reaction starts with U-238, and ends up with Pu-239. The intermediate products, U-239 and Np-239, are both radioactive with half-lives of, respectively, 23 minutes and 2.35 days.

A few reactors of this type have been built with a power of up to a couple of hundred megawatts. The experiences in France and England are promising, but breeder reactors, so far, are not economically viable. However, the prospects for improvements are good and this type of reactor is expected to be utilized more in the coming decades.

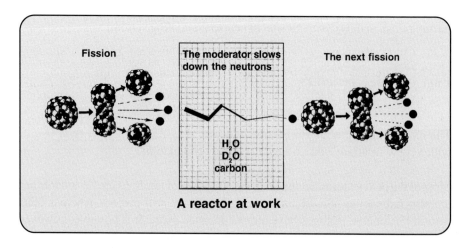

A fission results in the release of 2 to 3 fast neutrons, which after losing some of their energy, can take part in new fissions. A moderator must consist of atoms with a mass similar to the neutron and should not "eat" or absorb the neutrons. Heavy water (D_2O) is excellent but expensive; thus ordinary water and carbon are used extensively.

A natural reactor
The Oklo phenomenon

When Enrico Fermi started the first man-made reactor under Stagg Stadium in Chicago in 1942, everybody assumed that this was the first reactor. About 30 years later, French physicists discovered that nature itself had beaten Fermi by approximately 2 billion years. They reported that in Gabon in Africa there was, once upon a time, a *natural fission reactor* that operated for several thousands of years.

Oklo is in Gabon, close to the equator. In this area, regions rich in uranium have been found. In small "pockets", where the concentration of uranium is particularly high, fission of U-235 has occured in a similar manner as that for man-made reactors.

Some exciting work led to the discovery of the Oklo phenomenon. In tests taken from uranium mines, there is a very distinctive ratio between the isotopes U-238 and U-235. In the samples from Oklo the amount of U-235 was smaller than expected. In some samples, hardly more than 50% of that expected was found. The conclusion, after a long series of tests, was that U-235 had been "burned up" in the same way as in an ordinary fission reactor.

The uranium in Oklo was formed about 2 billion years ago. At that time the amount of U-235 was about 3% (today it is only 0.71%). Consequently, the amount of U-235 was approximately like that found in an enriched reactor.

The requirements for a reactor involve materials that can moderate (slow down) the neutrons formed. In Oklo the moderator was water bound to the minerals.

There were probably three "pockets" in which fission could occur. The effect was small, like a research reactor, and it was controlled by a variation in the water content. Fission products were formed, but the radioactive isotopes disappeared a long time ago. The stable end products can still be found.

A very interesting result, with regard to storage of radioactive waste, is that the fission products have not moved significantly in the course of 2 billion years.

The plutonium (Pu-239) formed in the Oklo-reactor was slowly transformed to U-235 (α-decay), which then went into the burning cycle. Today the occurrence of U-235 is too small (0.71%) for any new natural fission reactors. The Oklo-phenomenon will not reappear, but the uranium mines in Gabon have yielded a number of interesting facts.

The first man-made reactor

December 2, 1942, 4 years after the discovery of fission, the first reactor was started under the stands at Stagg stadium in Chicago.

Shortly after the discovery of fission, **N. Bohr** and **J.A. Wheeler** found U-235 (only 0.71 % of uranium consists of this isotope) to be a fissile atom.

The Hungarian physicist **Leo Szilard** had the idea about a chain reaction. He assumed that if the fission process released neutrons, they could, in turn, be used to fission other atoms. It was found that the fission process released 2 to 3 neutrons with high energy and speed (fast neutrons).

The most abundant isotope, U-238, is not easy to split. On the contrary, U-238 will absorb the high energy neutrons released, and thus stop the chain reaction. The absorption of slow neutrons is far less, and consequently, it became a goal to slow down the neutrons released in the fission. In a reactor it is necessary to mix in a moderator with the uranium core. A moderator consists of light atoms (preferably close to the weight of the neutron).

In ordinary water, the hydrogen atom has the right weight, but readily absorbs neutrons. Heavy water, containing deuterium is also a useful moderator and is used in certain reactors.

During the war, Norway was the only producer of heavy water and, since this country was occupied by the Germans, **Fermi** and co-workers had to use graphite as a moderator. They built a reactor in a squash court at Stagg stadium in Chicago. They built layer after layer of graphite. In the layers they left room for boxes of natural uranium (about 2.5 kg in each box). They had 10 control rods, made of cadmium which could absorb and control the neutrons.

If fission, on average, gave one neutron that could split a new atom, the process would go by itself. This *reproduction factor*, as Fermi called it, must be larger than 1.0. They measured the neutron flux all the time. The construction had the form of an ellipsoid (like an egg), 7.6 meters wide and about 6 meters high. It had 57 layers of graphite (385 tons) and about 40 tons of uranium oxide.

December 2, 1942 was a cold day with snow in Chicago. It was a tense atmosphere at Stagg stadium. Several people were gathered on the balcony where the counters were located (Szilard, Wigner, Compton, and others). Fermi gave the order to slowly remove the control rods and the neutron flux (measured by boron-trifluoride counters) increased. Finally he asked to take the last control rod 12 feet out. The clicking of the counters increased to a continuous roar. Fermi raised his hand and said: *"The pile has gone critical"*.

The reproduction value had been 1.0006. The first day the reactor operated for 4 minutes at an intensity of half a watt. Wigner, for the occasion, had bought a bottle of Italian Chianti and the successful experiment was celebrated.

AIP Emilio Segrè Visual Archives

Enrico Fermi

Nuclear Power and Radioactive Releases

Radioactive isotopes are produced in all types of reactors used for power production. Radioactive waste has many origins: from uranium mining; from the preparation and treatment of the fuel; from the normal operation of the reactor; and from the decommissioning of a nuclear power station. This radioactive waste may be in different forms such as the solid, liquid, or gaseous state. The activity is usually too high for a direct release to the environment and it is, therefore, necessary to use some type of isolation or storage.

According to international practice, radioactive waste is divided into the following categories:

1. Low-level radioactive waste

This type of waste contains isotopes with short half lives and/or the activity is low. Some of this waste may be released to the environment from the normal operation of reactors, some is stored until the radioactivity decays away, and some must be disposed of in a suitable facility.

2. Long-lived waste with medium level activity

This waste does not produce heat and comes from reprocessing and production of mixed fuel (plutonium/uranium). There are small volumes of this type of waste produced in normal reactor operations.

3. High-level waste

This type of waste consists of long-lived radioisotopes and comes from used fuel that is not reprocessed or from components separated in reprocessing.
The waste may be in the liquid or glassified form and the activity is high enough to produce heat. The glassified waste contains more than 99% of the total activity that was present before treatment.

Reprocessing

Used reactor fuel contains approximately 1% unused U-235, more than 90% of the original U-238, 0.5 to 1% Pu-239 and Pu-241, and small amounts of Np-237, as well as some other heavy elements with atomic numbers between 90 and 103.

The purpose of reprocessing is to separate the fissionable uranium and pluto-nium. Pu-239 can be used to make fuel for light water reactors, for fuel in a breeder reactor or for military purposes. The uranium can be enriched and used again.

A number of countries have built reprocessing plants: France (Marcoule, Cap la Hague), India (Tarapur, Trombay), Japan (Tokai Mura), Great Britain (Dounreay and Sellafield), Russia (Kyshtym) and USA (NFS, West Valley). Today France and Great Britain have the largest reprocessing capacities.

The reprocessing itself is done chemically. Since the fuel is highly radioactive the processes must take place in protected areas with remote controls and robots. When the fuel rods have been cut into pieces, they are dissolved in nitric acid. The waste is then divided into three groups:

1. **Uranium** (in nitrate form), which is purified and transferred into the oxide form (UO_2).

2. **Plutonium** (in nitrate form) is also purified and transferred into the oxide form (PuO_2).

3. **The fission products** are highly radioactive wastes in nitrate form (liquid) and are stored in steel containers. After a few years (up to 5 years maximum) the waste is converted to the solid state.

Several methods have appeared for transforming the high active liquid waste into the solid form. A method, introduced in France, involves heat treatment of the waste which in turn is transformed into a boron-silicate glass. The glassification prevents dispersion.

Treatment and Storage

The main goal of treatment and storage of radioactive waste is to bring the radioactivity into a form which is suitable for permanent storage. The purpose is to isolate the radioactive waste from the environment.

The low and medium level radioactive waste is largely transformed from liquid state into the solid phase. Small amounts of activity are released to the environment as *controlled releases*.

The normal procedure today is to store the solid waste for 5 to 10 years in the neighborhood of the reactor or reprocessing plant. In this way, most of the radioactivity decays before the waste is transported to other storage areas. The storage consists of the following two steps:

1. **Intermediate storage.** Storage of this type is not permanent. The waste which is stored can easily be retrieved. The storage needs security surveillance and control.

2. **Final storage.** The waste is put into permanent storage. It must, however, be technically possible to retrieve the waste. Final storage implies that continuous security surveillance is not necessary.

The requirements for control and storage of radioactive waste is far more strict than those for other industrial wastes. The final storage takes place in stable geological areas (salt, clay, granite), preferably 500 to 1,000 meters under ground. The storage place should give protection against external disturbances and act as a barrier against any dispersion of the radioactive isotopes. The storage places should be in areas that are not likely to be disturbed in the future by events such as a a new ice age.

With the requirements mentioned, the radioactive waste involves a risk that is comparable to the risk from natural radioactivity.

In 1985, the nuclear power industry produced about 500 cubic meters of highly radioactive waste. This represents, in volume, about a layer of 1 meter thick on a football field. The volume is, as you can see, negligible compared to the enormous amounts of waste from coal and other fossil fuels.

When the fuel is taken out of the reactor, the short-lived radioisotopes dominate. After some time, these isotopes decay and the waste is dominated by Cs-137 and Sr-90. After 20 years, these two isotopes constitute about 99.7% of the activity.

The types of radioisotopes determine the problems that may develop with the waste. Some isotopes can concentrate in the food chain. For example, Cs-137 is readily absorbed by reindeer lichen and can therefore enter animals that eat lichen. It is also known that iodine is enriched in sea plants and dairy products. If the iodine is ingested by a human, it will be concentrated in the thyroid gland. Iodine-131 has a half-life of only 8 days and it does not present a long-term storage problem.

The ICRP has estimated upper limits for acceptable concentrations of radioactive isotopes in air and water. These limits may be used to calculate an index for risk.

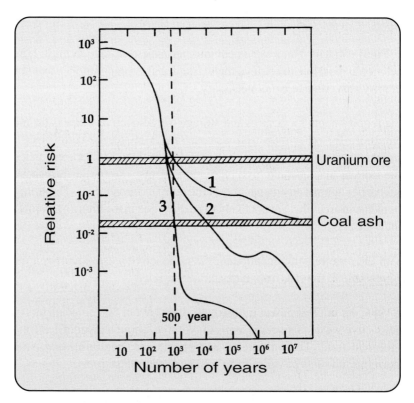

Figure 13.1. Risk index for long time storage of radioactive waste from a reactor. The curve marked 1 is for untreated waste from a reactor. Curve 2 represents the waste for which 1% plutonium and 0.1% uranium is removed and curve 3 represents waste with all the actinides removed. The horizontal bars represent the risk index for the radioactivity in coal ash and mined uranium. Courtesy of Finn Ingebretsen, Inst. of Physics, Univ. of Oslo

Waste from nuclear- and coal-power

Both nuclear power and fossil fuel power yield waste products. In the case of fossil fuel, large amounts of gases, mainly carbon dioxide (CO_2), are released directly to the atmosphere. These are gases that pose a potential problem with regard to the greenhouse effect. The regulations for release from nuclear installations are far more strict and there are plans and agreements for how to treat and isolate the waste.

For the years to come, there is no doubt that fossil fuel and nuclear power will be the two dominating sources of energy. It is of interest to compare the types and amounts of waste, and in order to do that, we use a 1000 MW power plant as a reference (data depends on the quality of the coal).

Nuclear power

Annual waste production:

1. High level active waste: 27 tons of used fuel. If reprocessed and glassified, the volume is about 3 cubic meters.

2. Medium level active waste: 310 tons.

3. Low level active waste: 460 tons.

4. Controlled release to the environment is without significance for health.

5. Products from the uranium mines produce a smaller volume than that for coal mining.

Coal fired plant

Annual waste production:

1. Carbon dioxide (CO_2): 6.5 million tons.

2. Sulfur dioxide (SO_2): 44, 000 tons.

3. Nitrogen oxides (NO_x): 22, 000 tons.

4. Ash: 320, 000 tons. Included is about 400 tons with toxic heavy metals like arsenic, cadmium, mercury and lead.

What can we do with all the waste?

Figure 13.1 gives the risk index for longtime storage of fission products and for the mined uranium.

Coal contains small amounts of radioactive isotopes, and the risk index for the activity in the coal ash is also given in Figure 13.1. The values are given using the assumption that the amount of power generated by a nuclear reactor is instead generated by a coal fired generator. After about a thousand years, the waste from a nuclear reactor has a smaller risk than that from a generator using coal.

Nuclear Power and Safety

The Chernobyl and the Three Mile Island accidents demonstrate that large nuclear accidents may occur. The two accidents also demonstrate the significance of having *reactor containment*. Without a good containment shell of reinforced concrete, the radioactive release to the environment was considerably higher at Chernobyl.

Calculations based on western reactors indicate the possibility of one major accident per 10,000 reactor years. On the other hand, for some of the Russian reactors without containment and with other weak points, there is a possibility for another accident in the next few years.

The goals for reactor safety are:

1. If the reactor core is destroyed the containment should remain intact. It should be possible to arrive at a stable situation where, for example, the damaged reactor is covered with water.

2. The stable situation should be reached with the release of no more than 0.1% of the Cs-134 and Cs-137 activity.

3. Even though a major accident with a release of large quantities of radioactivity can not be completely ruled out, the safety systems should be designed in such a way that the probability of an accident is extremely small.

Reactor safety

The power plants for the next century will be powered mainly by fossil fuel and nuclear fuel. Other energy sources either lack capacity or the technology is a long way off. All energy sources are combined with a certain risk for accidents. It is, therefore, very important to have acceptable safety practices for the different power plants. For nuclear power, it is a primary goal to prevent damage to the reactor core. The safety regulations have to be organized as a "defense in depth". If an accident should occur, the defense system should be able to reduce the consequences and prevent the release of radioactivity.

The barrier principle

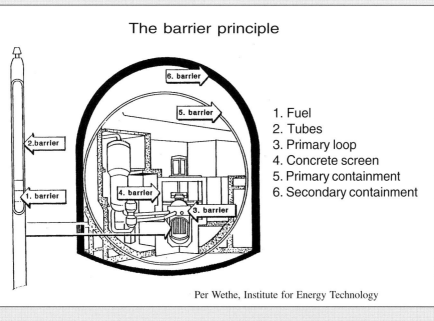

1. Fuel
2. Tubes
3. Primary loop
4. Concrete screen
5. Primary containment
6. Secondary containment

Per Wethe, Institute for Energy Technology

The figure indicates a model for the safety system. *It is very important to have good reactor containment.* The primary safety functions are:

1. Control of the effect.
2. Cooling of the fuel.
3. Prevention of toxic release.

The goal is to *minimize human failures* and to make the *safety functions automatic.*

Safety requirements include competence, training, inspections, and testing of equipment. Furthermore, safety analyses must be performed in order to see how the reactor behaves in different situations.

Radioactivity in the ocean

Most of the radioactive isotopes in the ocean are from natural sources. A small fraction comes from nuclear tests in the atmosphere and the nuclear industry. The radioactivity varies from one region of the ocean to another and this variation is due to the differences in the natural sources. Regions with high natural radioactivity are the Eastern Mediterranean, the Red Sea and the Persian Gulf.

NATURAL ISOTOPES			ARITIFICIAL ISOTOPES	
Isotope	Bq/liter	In percent of tot. activity	Isotope	Bq/liter
K-40	$1.2 \cdot 10$	96.0	**H-3**	10^{-2} - 2.7
Rb-87	$1.1 \cdot 10^{-1}$	0.9	**C-14**	$0 - 1.5 \cdot 10^{-3}$
U-234	$4.7 \cdot 10^{-2}$	0.4	**Cs-137**	$7 \cdot 10^{-4}$ - $2 \cdot 10^{-1}$
U-238	$4.1 \cdot 10^{-2}$	0.3	**Sr-90**	$4 \cdot 10^{-4}$ - $1 \cdot 10^{-1}$
Pb-210	$5.0 \cdot 10^{-3}$	0.04	**Pu-239**	$7 \cdot 10^{-6}$ - $3 \cdot 10^{-4}$
Po-210	$3.7 \cdot 10^{-3}$	0.03		
Ra-226	$3.6 \cdot 10^{-3}$	0.03		
U-235	$1.9 \cdot 10^{-3}$	0.02		

The table demonstrates that K-40 is responsible for 96% of the radioactivity in the ocean. This isotope will enter all living organisms in the sea. The uptake of K-40 in the different types of fish is not well known. Some data have been presented for the concentration factor (see Table 13.1). According to the table above, and using a concentration factor of 16, the fish should have an activity of about 190 Bq/kg (for humans the concentration is smaller, from 40 to 70 Bq/kg). As a curiosity, it can be mentioned that the calculated annual natural dose to fish in the ocean seems to be much smaller (about 0.5 mSv) than that obtained by animals on land!

The extra dose from anthropogenic isotopes is, so far, negligible. Furthermore, it can be concluded that fish contain a smaller amount of radioactivity than the meat from animals on land.

Radioactivity in the Ocean

Over the years there have been releases of man-made radioactive isotopes into the oceans. This is mainly due to fallout from the nuclear tests performed in the atmosphere and releases from the reprocessing plants in Sellafield, England and La Hague in France. There have also been some releases of low and medium level radioactive waste from research reactors and nuclear powered ships. Studies have been carried out in order to map the pollution and to measure the activity in sea water, in marine organisms and in sediments.

It is important to realize that the oceans contain natural radioactive isotopes such as K-40, U-238 and Ra-226. The radioactivity in sea water has an average concentration of 12.5 Bq per liter. K-40 is responsible for 96% of this activity. The man-made pollution from nuclear tests and other releases has, on average, resulted in an increase of 1 Bq per liter.

By model calculations and measurements, it is possible to see how man-made releases follow the ocean currents. One example is given in Figure 13.2. Here we see the relative amounts and the time scale for dispersion of the release from Sellafield in England. It appears that the release of Cs-137 can be observed in the West-Spitzbergen current after 4 to 6 years and in the East-Greenland current after 6 to 8 years.

Plutonium is usually bound to sediments and inorganic materials and the dispersion is restricted to a few km from the place of release.

Radiation Doses to Fish

Discussions have recently taken place about the release of radioactivity to the northern part of the ocean (North Atlantic, The Barents Sea, the sea north of Russia) because of the nuclear tests on Novaja Zemlja, the Russian marine fleet and nuclear installations on Kola. Furthermore, the possibility for transport of radioactive waste via the large north-bound Russian rivers has been mentioned. All this has been characterized by environmental organizations as a threat to the environment and to the fish living in these waters. It has been suggested that large efforts should be made in order to prevent further dispersion of the radioactivity before the fishing industry has a serious problem.

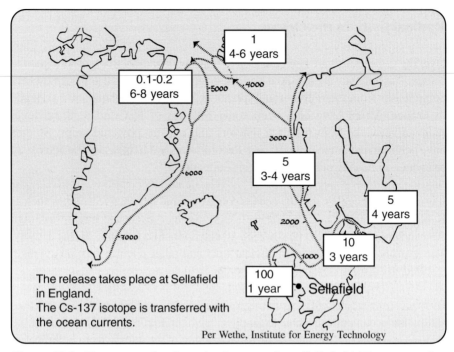

Figure 13.2. Transport of radioactive isotopes from Sellafield. The top number in the small white boxes is the relative concentration of Cs-137 in Bq/m³ and the number underneath it is the transport time in years. The distance from Sellafield in km is given on the dotted curves.

Is it possible for us, with the knowledge now available, to say something about doses to fish and to the people who eat fish? Can any quantitative data be given about the threat?

Doses to fish would be extremely small and would not harm fisheries. The doses would be a sum of external and internal radiation. Since water is an excellent absorber of γ-radiation, the external dose to fish is negligible. The radioactive isotopes are incorporated into the food chain, starting with plankton. It is known that a number of elements can be concentrated in fish. The concentration factor (called CF) is defined in the following way:

$$CF = \frac{C_{Fish}}{C_{Water}}$$

Here C_{Fish} is the concentration of the element/compound in question in fish and C_{Water} is the concentration of the same element/compound in water.

Knowledge is limited about the different concentration factors. A few experiments have been done, but the results are uncertain. Some values are presented in Table 13.1. There is no information about the variation from one type of fish to another. It must be assumed that there are large differences since the food chains are different. The data in Table 13.1 yield some interesting differences between fresh water fish and fish from the sea, which is probably due to differences in the food chains.

Table 13.1. Concentration factors in fish

Isotope	CF Fresh Water	CF Sea Water
K-40	4,400	16
Cs-137	3,680	48
Sr-90	14	0.43

For environmental purposes it would be of interest to gain more information about the concentration factors for both natural and artifical radioactive isotopes. As mentioned above, the concentration of radioactive isotopes in the sea is completely dominated by the natural isotope K-40. Among the artificial isotopes released to the ocean, Cs-137 is the most important. Today the concentration of Cs-137 is from 100 to 17,000 times smaller than the K-40 concentration.

The amount of plutonium in the ocean is very small. This isotope is mainly bound to sediments. Plutonium emits α-particles with a range of less than 1.0 mm in water. Most of the plutonium that enters an organism via food and water will be rapidly excreted. The ICRP assumes that only 30 ppm (parts per million) is absorbed in the blood for humans. In humans, the important path of entry of plutonium is by inhalation but since fish don't have lungs, this problem is absent in fish. This means that plutonium should not be a significant problem for the health of the fish population nor a concern when harvesting food from the sea.

Submarines and pollution

Almost all submarines are now powered by nuclear reactors. Consequently, an accident involves a risk of radioactive pollution. On April 7, 1989, the Russian submarine Komsomolets sank near Bear Island in the North Atlantic. The boat with reactor, fission products and war heads, is now at a depth of 1,680 meters. The question is: will the radioactivity in this submarine present a threat to the environment?

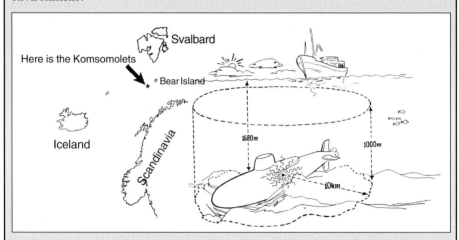

According to Russian documents, the submarine contained isotopes with an activity of 25,000 TBq (680,000 Ci) in 1989. There were about 2,800 TBq Sr-90 and 3,100 TBq Cs-137. In addition, there were about 10 kg of plutonium, including that in the war heads.

Sr-90 and its decay product Y-90 emit only β-particles. Cs-137, on the other hand, emits both a β- particle and γ-radiation (decay scheme in Figure 2.4). The only biological effect due to these important fission products would occur if the isotopes in some way got into the food chain.

Let us assume a major leak with dispersion of the activity to the surrounding water.

If the activity is distributed evenly in the water within a cylinder with a radius of 10 km and a height of 1000 meters (insert above shows a diagram of rather small volume) the activity in 1989 would have been 8.9 Bq/l in the case of Cs-137 and 9.8 Bq/l for Sr-90. The activity decreases with a half-life of about 30 years for both isotopes.

If we assume concentration factors such as those given in Table 13.1, the activity in the fish in the area would be 3.8 Bq/kg of Sr-90 and 470 Bq/kg of Cs-137.

It can safely be concluded that a leak from the sunken submarine will not pollute the food chain and would not be harmful to people that eat the fish.

Plutonium and Chemical Compounds

Throughout this book an attempt has been made to point out that radiation is of significance to our health. The emphasis, so far, has been placed on detrimental effects, but the positive attributes of radiation and radioactivity have also been discussed. Because of the negative effects, radioactive isotopes have been called poisonous or toxic. The toxicity is related to the types of radiations emitted.

The nuclear power industry is the largest source of radioactive isotopes but plays only a minor role in delivering radiation doses to humans. The medical uses of radiation along with natural background radiation deliver the largest doses.

A number of chemical compounds are dispersed to the environment accidentally or sometimes deliberately. Some of these compounds are very hard to break down. It is, therefore, difficult to rank the toxicity of the different compounds. One of the major problems is the shape of the dose–effect curves at small doses. Environmental problems created by chemical compounds that are *not* radioactive are often similar to those that are radioactive.

Plutonium has sometimes been classified as the most dangerous element in the world. If plutonium is concentrated into one lump, large enough to become critical, it represents a serious threat. The fact is that, when plutonium is thinly dispersed around the globe, it is not dangerous.

Pu-239 emits α-particles. This means that plutonium does not represent any radiation problem when it is outside the body. The range of α-particles in air is only a few cm. In tissue the range is only a few cells but the ionization density is high, a characteristic of high-LET radiation. Plutonium can enter the body via two routes:

1. Consumption of food. The plutonium received in food will mainly be excreted from the body. The uptake from the intestine into the blood stream is small (see above). Consequently, the plutonium in food is of minor importance.

2. Inhalation. Inhalation of air containing plutonium can lead to plutonium in the lungs and the bronchial tubes. How this isotope is distributed in the respiratory tract and the rest of the body is a function of the size of the particulates and their chemical state. Some of the plutonium will remain in the respiratory tract with a short half-life, some with a long half-life. The more water-soluble the plutonium compound, the more of the isotope that will dissolve in lung fluids and be taken up by the circulating blood. Of that which is taken up by the blood, 45% will be deposited in bone with a biological half-life of 100 years and 45% in the liver where it will have a half-life of 40 years. A very small percentage is taken up by the gonads where it will remain indefinitely.

The conclusion is that it is inhaled plutonium that delivers damage to the body. The most important long-term effect is the risk of cancer (cancer of the lungs, liver and bone). Experiments have been carried out with dogs, rats and rabbits that have breathed air containing plutonium. The doses involved were very large and resulted in lung cancer.

These small animal experiments gave the very important finding that the latent period depends on the dose. The smaller the dose, the longer the latent period. If (from these experiments) the *dose–latent period curve* is extrapolated down to a dose region that may occur in a human population exposed to a plutonium accident, the latent period would be significantly longer than the life expectancy of a human.

Plutonium and the environment

In the Chernobyl accident there was a release of plutonium with a fallout that was concentrated in a region within 30 km of the reactor. Under weather conditions with strong winds, this plutonium dust could be picked up into the air and consequently, present a radiation risk.

Another source for plutonium pollution comes from the many nuclear tests performed in the atmosphere in the 1960s. A certain fraction of the plutonium was not fissioned and resulted in fallout that is assumed to be about 6 tons altogether. If we assume that this plutonium is distributed evenly around the world, over land and sea, this amount of plutonium represents a pollution of approximately 26 Bq/m^2. Since most of the atmospheric tests were perfomed on the northern hemisphere, the plutonium pollution in that hemishpere may be up to 50 Bq/m^2.

With regard to plutonium, the following can be concluded:

1. Plutonium dispersed into the environment represents a relatively small or nonexistent health problem.

2. Plutonium material *gathered in a lump* that can be made critical is extremely dangerous.

3. Plutonium used in reactors can be an important energy resource that does not contribute greenhouse gases to the environment.

 Our societies have the possibility to decide how to use plutonium, either safely or unsafely. Plutonium is a valuable energy source for the world.

Remedial Action

In some countries, resources and monies are available for environmental actions. It is important to judge the different proposals for action carefully in order to use resources in the best way. A cost-benefit analysis should always be made.

In the case of radiation, large amounts of money have been spent to reduce the radioactivity in meat. It is known that animals feeding on grass or lichen in polluted areas will take up this activity. For example, Cs-137 was taken up by sheep and reindeer after the Chernobyl accident. The radioactivity reached (in some parts of Scandinavia) several thousand becquerel per kilogram (for the reindeer it was measured up to 100,000 Bq/kg).

Since the biological half-life for Cs-137 in sheep is of the order of 3 weeks, it is possible to feed the animals for a few weeks on nonradioactive food before slaughtering. Considerable amounts of money have been spent in these feeding actions. The radioactivity in the meat was reduced and a threshold limit of 600 Bq/kg was set for selling the meat.

A critical question was not addressed. What is the cost-benefit of this remedial action? The authorities thought the benefits justified the cost, but from a radiobiology point of view, this was a waste of money. The extra radiation doses involved were already very small, even without any remedial action. The only argument that favors this remedial action is a psychological one. People were given the impression that authorities took action on behalf of their safety.

Closing Remarks

The purpose of this book is to provide information about ionizing radiation, its use and its consequences. We have provided information about the applications of radiation that benefit society and the different parameters that must be considered for the use of radiation technology in medicine, research, and industry.

The risk factors for radiation are easily identified making it possible to establish rules and regulations to limit radiation doses, both to the individual and to society.

It is hoped that the readers have gained sufficient knowledge to carry out some dose calculations and to compare the results with the doses from natural background radiation.

Some examples of dose calculations are given in Chapter 14. Furthermore, references, where additonal information can be found, are given in the last section.

Chapter 14

Exercises, Examples and Scenarios

Those interested in radiation should have, after reading this book, the knowledge to undertake dose calculations and to judge the size and extent of possible radioactive pollution. In order to perform calculations about radioactivity, we will use the equations presented in Chapter 3. If you can use the simple equations 3.1, 3.2 and 3.3 you can embark on all kinds of estimates.

The reader is invited to perform the exercises presented on the following pages. You can look at the answers and follow the explanations given but you should try first.

Exercise 1. Carbon-14 and number of atoms for 1 Bq

It was noted in Chapter 3 that the C-14 activity can be used to determine the age of artifacts. Two types of measuring techniques were discussed; either to determine the number of disintegrations (number of Bq) in a sample or to calculate the total number of C-14 atoms in the sample. The latter method is far more sensitive because 260.7 billion C-14 atoms are needed to give one Bq.

The first exercise proves this.

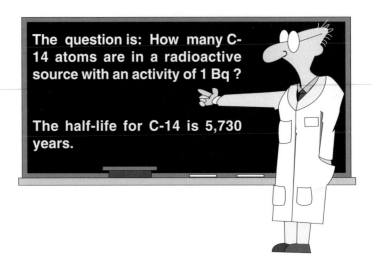

The question is: How many C-14 atoms are in a radioactive source with an activity of 1 Bq ?

The half-life for C-14 is 5,730 years.

Solution

You can solve this question by starting with the equation 3.1 for the activity (i.e. the number of disintegrations per second):

$$dN/dt = N\lambda$$

Here N is the number of atoms in the source (the answer to our question). Furthermore, $dN/dt = 1$ Bq. The disintegration constant λ in $\sec^{-1}$ is, according to equation (3.3) in Chapter 3, given by:

$$\lambda = \frac{0.693}{t_{1/2}} = \frac{0.693}{5730 \cdot 365 \cdot 24 \cdot 60 \cdot 60}$$

As you can see the half-life, which was given in years, is transformed to seconds because the activity is given in Bq. It should now be easy to find that: $N = 1/\lambda = 260.7$ billion atoms.

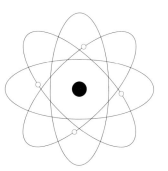

Exercise 2. Ra-226 and I-131

a) The disintegration constant for radium (the isotope Ra-226) is $1.4 \cdot 10^{-11}$ s^{-1}. How many radium atoms disintegrate per second in a 1 mg sample of radium ? Or put another way, what is the activity (in Bq) of a 1 mg source of radium?

Avogadro's number (number of atoms in one mole) is $6.023 \cdot 10^{23}$.

b) The iodine isotope I-131 has a half life of 8.04 days. What is the activity (in Bq) in a 1 mg sample of I-131 ?

Answers: a) $3.7 \cdot 10^7$ Bq and b) $4.59 \cdot 10^{12}$ Bq.

You solve this question by starting out with the same equation as that used in exercise 1. You need one additional piece of information; namely, Avogadro's number. By definition, Avogadro's number is the number of atoms in a gram of the specified molecule. For Ra-226 a gram of molecule is 226 grams, whereas for I-131 it is 131 grams.

The answer to the first part of this exercise can be found directly from the definition of the Curie unit (see Chapter 4). 1 mg radium has an activity of 1 mCi.

Note the large difference in activity of the two sources, both weighing 1 mg. The reason is, of course, the large difference in half-lives (or disintegration constant) for the two samples.

Note! The half-lives (or disintegration constants) are given in two different forms in this exercise.

Exercise 3.

The isotope Co-60 may be made in a reactor using the following reaction:

$$^{59}Co + n \Rightarrow {}^{60}Co + \gamma$$

Here n is a neutron. The half-life for Co-60 is 5.3 years. The neutron irradiation is continuous until the activity is $3.7 \cdot 10^{10}$ becquerel (1 curie).
How many Co-60 atoms are formed in the sample when the irradiation is done?

Answer: $8.9 \cdot 10^{18}$ atoms.

Exercise 4. Carbon-14 dating

Some archeologists found a piece of wood which they assumed could be from a Viking ship. In order to find out more about this hypothesis they decided to determine the age of the piece of wood by C-14 analysis.

All living organic materials contain C-14 at a concentration of 15.4 disintegrations per minute per gram of pure carbon.

The piece found by the archeologists weighed 2 gram. The activity was 11.8 disintegrations per minute. The carbon content of the wood was 44%. How old was the piece of wood?

Answer: 1,144 years.

Exercise 5. A scenario where isotopes are used.

An iron ring used in one of San Francisco's cable cars weighs 50 grams. It is of interest to know how this ring is worn down in the engine. Radioactive labeling may be used to determine this.

The iron ring is irradiated in a reactor and the radioactive isotope Fe-59 is formed. This isotope has a half life of 45 days. The activity in the ring at the end of the irradiation is $3.7 \cdot 10^5$ Bq. After 40 days use, a small fraction of the ring has been worn off and radioactivity can be found in the lubricating oil.

100 ml of the lubricating oil is measured and was found to have an activity of 14 disintegations per minute. Altogether 3 liters of lubricating oil are in contact with the iron ring. How much of the iron ring has been worn down in 40 days?

Answer: 1.75 mg.

In this exercise you can note that the activity is given both in Bq as well as in disintegrations per minute. You must be careful to use the same units.

Let us assume that P grams of iron are worn off and are present in the 3 liters of lubricating oil. The P gram has an activity of $(14/60) \cdot 30 = 7$ Bq. The activity decays according to equation 3.1, which yields the following equation:

$$7 = \frac{P}{50} \cdot 3.7 \cdot 10^5 \cdot e^{-\frac{\ln 2}{45} \cdot 40}$$

This equation can be solved for P.

Exercise 6. Cesium from Chernobyl

The most important isotope released in the Chernobyl accident was Cs-137, with a half-life of 30 years.

The amount released in the accident was given as 38,000 TBq (1 TBq is 10^{12} Bq). What is the weight of the Cs-137 released in the accident?

(Avogadro's number is: $6.023 \cdot 10^{23}$).

Answer: 11.8 kg.

You can calculate this by using the same equation as used in exercise 1.

The reactor and the surrounding area three days after the accident occurred.

Exercise 7. A scenario with pollution by isotopes

In a research laboratory, work is going on with the radioactive isotope Na-24. One day an accident occurred resulting in contamination of the laboratory. The radiation authorities found that the activity was 100 times that acceptable. They decided to close the laboratory until the activity reached an acceptable level.

Na-24 has a half-life of 15 hours. For how long a time must the laboratory be closed ?

Answer: 100 hours.

Exercise 8. A scenario with radioactive food

Assume that you are invited to a dinner and your host tells you that you will be served reindeer meat containing Cs-137 with a concentration of 10,000 Bq/kg. (This was far more than the threshold limit set for meat in most European countries after the Chernobyl accident).

Before you accept that invitation you would like to make a rough calculation of the radiation dose associated with this particular dinner.

You weigh 60 kg and you eat 200 grams of reindeer meat (or 2,000 Bq of Cs-137). How large is the total dose from this dinner?

Hints: Use the decay scheme shown in Figure 2.4 and remember that $1\,eV = 1.6 \cdot 10^{-19}\,J$.

Answer: 0.03 mSv (mainly in the course of one year).

Help with the solution

Cs-137 has a physical half life of 30 years, but is rapidly excreted from the body. Assume that the biological half-life is 3 months. According to equation (3.5) this gives an effective half-life of 90 days.

Both the number of Cs-137 atoms and its activity (A), vary with time as that given in equation 3.2. The total number of disintegrations is given by:

$$x = \int_{0}^{\infty} A_o e^{-\lambda t} dt = \frac{A_o}{\lambda} = 2.24 \cdot 10^{10}$$

Here A_o = 2,000 Bq, which is the amount of radioactivity you ate during the meal (t = 0).

You can notice that the integration time goes to infinity but during the first year more than 93% or the main part of the dose is received.

Look at the decay scheme in Figure 2.4. All β-particles emitted will be absorbed in the body. The average β-energy is approximately 1/3 of the maximum energy given in the decay scheme. This means that the β-particles contribute to the energy absorption with 0.2 MeV per disintegration. The γ-radiation will be partly absorbed in the body and partly exit the body. It is assumed that about half of the γ-energy is deposited in the body. Altogether, it is reasonable to assume that every disintegration yields an energy absorption of about 0.5 MeV (see also Chapter 9).

We assume that cesium is distributed evenly throughout the body giving us a total energy deposition in the body (weighing 60 kg) of $1.12 \cdot 10^{16}$ eV. Since 1 eV = $1.6 \cdot 10^{-19}$ J, the following dose is obtained:

$$D = \frac{1.12 \cdot 10^{16} \cdot 1.6 \cdot 10^{-19}}{60} = 3.0 \cdot 10^{-5} \, \text{J} / \text{kg}$$

$$D = 0.03 \text{ mGy}$$

Since the radiation consists of β-particles and γ-radiation and the weighting factor is 1, the biological effective equivalent dose is 0.03 mSv.

You should compare this dose with the "normal" annual dose, which is more than 100 times larger. You can also compare the "dinner dose" to the doses you may receive when flying.

Exercise 9. A swim in polluted water

(A more difficult exercise)

Now we will consider a scenario that may surprise you. We will take all the Cs-137 from the Chernobyl accident, which was spread out over the whole world (in exercise 6 we found that it was 11.8 kg or 38,000 TBq) and pour it into a rather small lake. The lake is 10 km by 10 km in area and 20 meters deep. Now, assume that all the cesium is mixed evenly in the water and nothing settles out.

Question: What would the radiation dose be if you took a 10 minute swim in this lake?

Answer: Approximately 3.5 µGy or 3.5 µSv

The dose is extremely low and you may find it hard to believe unless you have made your own calculations. In fact, if you reduce the lake to 1 km by 1 km (about the size of a large "swimming pool" – a small pond) the dose would increase to 0.35 mGy, which would be the dose you obtain in a chest x-ray.

Before you look at the solution of the question you should try to find the answer yourself.

Solution: First of all, we assume that you do not drink the water when you are swimming.

We start by calculating the activity in the water when 38,000 TBq Cs-137 is distributed throughout the lake. The artificial lake of 10 km by 10 km and with a depth of 20 meter contains $2 \cdot 10^{12}$ liter of water. If a source of $38 \cdot 10^{15}$ Bq is mixed evenly into the lake, the activity would be 19,000 Bq/l.

Since no drinking takes place during the swim, we will concentrate only on the external radiation.

The decay scheme for Cs-137 (Figure 2.4) demonstrates that two types of β-particles can be emitted. The majority (94.5 %) have a maximum energy of 0.512 MeV, and an average energy of approximately 0.2 MeV. In water, β-particles with an energy of 0.2 Mev would have a range of approximately 1 mm. This means that it would be only β-particles from a water layer of 1 mm around your body that reaches your skin. Collecting the 1 mm layer of water surrounding your body into one container gives approximately 1 liter of water. The β-particles are emitted in all directions and we assume that half of them hit the body. Your skin would be exposed to approximately 5.7 million β-particles during the swim of 10 minutes. The β-particles lose some energy before they hit the skin, reducing the average energy to about 0.1 MeV. If we assume that all this energy is deposited in the epidermis (about 0.1 mm thick, weighing 0.1 kg for the total body) the dose would be about:

$$D = 5.7 \cdot 10^6 \cdot \frac{0.1 \cdot 10^6 \cdot 10^{-19}}{10^{-1}} \approx 1 \cdot 10^{-6} \text{J} / \text{kg} = 1\mu\text{Gy}$$

Since the β-particles deposit their energy to depths of more than 0.1 mm the dose to the epidermis would be somewhat smaller. However, the calculation gives you an estimate of the skin dose from the β-particles.

The β-particles yield a negligible contribution to the total body dose. The dose to the body is, therefore, dominated by the γ-radiation. The γ-radiation from Cs-137 has an energy of 0.662 MeV. Both x-rays and γ-rays are absorbed easily in water (described by an exponential function). A layer of water of less than 10 cm will reduce the radiation from Cs-137 by 50%. This means that it takes 5 such "half-value layers" (50 cm of water) to reduce the radiation by 97%. Consequently, only the Cs-137 atoms within a distance of about 50 cm give you a significant dose when you are in the water.

Assume that your body has the shape of a cylinder, 180 cm high and 22 cm in diameter (this gives a weight of about 70 kg). The amount of water around you that can give you a radiation dose has the form of a cylindrical shell with a diameter of 1.22 meter, containing approximately 2000 liters of water with a total activity of about $4 \cdot 10^7$ Bq.

Since the radiation is emitted in all directions, approximately half of it is directed at you. The radiation will partly be absorbed before it reaches you. For a rough estimate of the dose, we divide the water into 5 layers around you. The thickness of each water layer is 10 cm (equal to one half-value layer). The amount of water in each layer will gradually increase as you calculate the volumes of consecutively larger shells: 181 - 294 - 407 - 520 - 633 liter.

Approximately 71% of the γ-radiation from the first layer, which is directed against you (half of the photons), will hit your body. For the next layer 35% will reach you, and then for layer 3 about 18%, layer 4 about 9% whereas only 4% from the outermost layer will reach you. (You can calculate this using an exponential function.)

Based on the above conditions you will find the γ-photons from $3.6 \cdot 10^6$ disintegrations hit you each second. For a swim of 10 minutes this gives about $2.2 \cdot 10^9$ *γ-photons striking you.* If we assume that all the energy from these photons is absorbed evenly in your body (70 kg) the radiation dose will be:

$$D = \frac{2.2 \cdot 10^9 \cdot 0.662 \cdot 10^6 \,\text{eV} \cdot 1.6 \cdot 10^{-19} \,\text{J} / \text{eV}}{70 \,\text{kg}}$$

$$D = 3.3 \cdot 10^{-6} \,\text{J/kg} = 3.3 \,\mu\text{Gy}$$

Conclusion

The radiation dose of 3.3 μGy, or 3.3 μSv, is surprisingly small. The result shows that the absorption of radiation by the water has a large protective impact.

You would, of course, have a different scenario if you started to drink the water or eat fish from the lake.

Exercise 10. I-131 and doses to the thyroid gland

In the regions around Chernobyl a large increase in the number of thyroid cancers has been observed. We assume that this is due the radioactive iodine released during the accident. We have very little information about the doses from I-131 but we can try to give a scenario and estimate the doses.

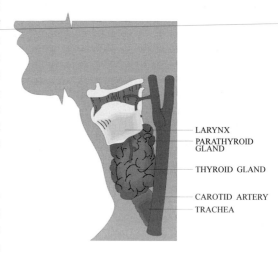

LARYNX
PARATHYROID GLAND

THYROID GLAND

CAROTID ARTERY
TRACHEA

Let us assume that I-131 entered the milk supply. I-131 has a half-life of only 8 days. This means that after 20 weeks the radioactivity is reduced to 10^{-6} of the initial value. Let us calculate the dose to the thyroid.

Question: What is the dose to the thyroid in the following scenario?

1. You drink 1/2 liter of milk every day in a period of 20 weeks.

2. The milk containing I-131 has a radioactivity level 1000 Bq/l.

3. All I-131 ends up in the thyroid gland. The biological half-life is very long so we assume that the "effective half-life" is equal to the physical half-life (8 days).

4. The thyroid weighs 25 grams.

The decay scheme for I-131 shows that the isotope emits a β-particle with maximum energy of 0.6 MeV and γ-radiation with an energy of 0.36 MeV.

Answer: 180 mGy

We assume that all β-particles, with an average energy of 1/3 of the maximum energy, are absorbed in the thyroid. A fraction of the γ-radiation escapes from the body (this is why this isotope is used for diagnostic purposes). Let us assume that 50 % is absorbed by the body. This results in each disintegration depositing about 0.4 MeV in the body. For simplicity, let us assume that all of this energy is deposited in the thyroid gland.

This scenario assumes you drink 70 liters of milk containing radioactive iodine (1000 Bq/l). All the radioactivity (70 kBq) is concentrated in a thyroid gland of 25 gram. The half-life is 8 days and the energy per disintegration is 0.4 MeV. The total number of disintegrations (X) is:

$$X = A_o/\lambda = 70,000/\lambda = 6.98 \cdot 10^{10}$$

The energy deposition in the thyroid is then:

$$Energy = 0.4 \cdot 10^6 \text{ eV} \cdot 6.98 \cdot 10^{10} = 2.8 \cdot 10^{16} \text{ eV}$$

The dose to the thyroid gland, weighing 25 gram is:

$$D = \frac{2.8 \cdot 10^{16} \cdot 1.6 \cdot 10^{-19}}{0.025} \approx 1.8 \cdot 10^{-1} \text{J / kg} = 0.18 \text{Gy}$$

Exercise 11. Biological half-life for K-40 in humans

The largest natural contribution to internal radiation dose is due to K-40. The level of K-40 is rather constant even though it varies with age and sex. In this exercise, we assume that the total body contains a constant level of 5000 Bq (about 70 Bq/kg). Furthermore, we assume that the daily consumption of potassium is approximately 2.5 gram. Since 0.0118% of the potassium consists of the radioactive isotope K-40, you can calculate that each day you eat about 76 Bq of K-40.

1. Carry out this calculation.

2. Calculate the biological half-life for K-40 if you eat 76 Bq per day and the level in your body is constant at 5,000 Bq

Physical half-life for K-40 is 1.27 billion years.
Avogadro's number: $6.023 \cdot 10^{23}$

Solution: The number of K-40 atoms per day in the food (N) is:

$$N = \frac{2.5 \cdot 0.0118}{100} \cdot \frac{6.023 \cdot 10^{23}}{40} \approx 4.4 \cdot 10^{18}$$

The number of Bq is therefore:

$$\frac{dN}{dt} = N\lambda = 4.4 \cdot 10^{18} \frac{\ln 2}{1.27 \cdot 10^{9} \cdot 365 \cdot 24 \cdot 60 \cdot 60} = 76$$

In order to calculate the biological half-life we use the same equation as above:

$$\frac{dA}{dt} = 76 = A\lambda = 5000 \cdot \frac{\ln 2}{t_{1/2}}$$

From this $t_{1/2} = 45$ days.

Additional Reading

1. *Radiation and Life*, Eric J. Hall, Pergamon Press, 1984.
2. *Environmental Radioactivity*, Merril Eisenbud, Academic Press, 1987.
3. *1990 Recommendations of the International Commission on Radiological Protection*, ICRP Publication 60, Pergamon Press, 1991.
4. *Annual Limits on Intake of Radionuclides by Workers Based on the 1990 Recommendations*, ICRP Publication 61, Pergamon Press, 1991.
5. *Radon and Its Decay Products in Indoor Air*, Editors: W. W. Nazaroff and A. V. Nero, John Wiley, 1988.
6. *Radioisotopic Methods for Biological and Medical Research*, Editor: Herman W. Knoche, Oxford University Press, 1991.
7. *Biological Radiation Effects*, Editor: J. Kiefer, Springer-Verlag, 1990.
8. *Sources, Effects, and Risks of Ionizing Radiation*, UNSCEAR Report, United Nations, New York, 1988.
9. *Atoms, Radiation, and Radiation Protection*, Editor: James E. Turner, John Wiley and Sons, Inc. (second edition), 1995.
10. *Radiation Biophysics*, Editor: Edward L. Alpen, Academic Press (second edition), 1997.
11. *Basic Clinical Radiobiology*, G. Gordon Steel, Arnold (second edition), 1997.
12. *Radioactivity and Health – A History*, J. Newell Standard, Office of Scientifical and Technical Information, Battelle Memorial Institute, 1988.
13. *A History of X-rays and Radium: with a Chapter on Radiation Units, 1895–1937*, Richard F. Mould, ICP Building & Contract Journals Ltd., London, England, 1980.
14. *History of Physics*, Spencer R.Weart and Melba Phillips, American Institute of Physics, New York, N.Y., 1985.
15. *Health Effects of Low-level Radiation*, Sohei Kondo, Atomic Energy Research Institute, Kinki University, Kinki University Press, Osaka, Japan Medical Physics Publishing, Madison, WI USA, 1993.
16. *A Century of X-rays and Radioactivity in Medicine with Special Reference to Photographs of the Early Years*, Richard F. Mould, Institute of Physics Publishing, Bristol, 1993.
17. *Radiation Biology*, Allison Casarett, Prentice-Hall Inc., Englewood Cliffs, NJ, 1968.
18. *The Children of the Atomic Bomb Survivors: A Genetic Study*, J. V. Neel and W. J. Schull, National Academy Press, Washington, D.C., 1991.

19. *The International Chernobyl Project. An Overview*. IAEA, Vienna. ISBS 92-0-1291-0, 1991.

20. *The International Chernobyl Project. Assessment of Radiological Consequences and Evaluation of the Protective Measures*. ISBN 92-0-129091-8. IAEA, Vienna, 1991.

21. *IAEA-report: Summary Report on the Post-Accident*, Safety Series No. 75-INSAG-1.IAEA, Vienna, 1991.

22. *Chernobyl Conclusions – International Conference on Radiation and Health*, Beer Sheva, Israel, 3.–7. November 1996. Conference supported by WHOand IAEA.

23. *Chernobyl Record*, Richard F. Mould, Institute of Physics Publishing, Bristol, UK & Philadelphia, USA, 2000.

24. *Chernobyl – The Real Story*, Richard F. Mould, Pergamon Press, London, 1988.

References

O. T. Avery, C. M. MacLeod, and M. McCarty, Studies on the chemical nature of the substance inducing transformation of pneumonococcal types; induction of transformation by a deoxyribonucleic acid fraction isolated from pneumococcus type III. *J. Experimental Medicine* **79**, 137–158 (1944).

M. J. Crick and G. S. Linsley, An assessment of the radiobiological impact of the Windscale reactor fire. *Int. J. Radiation Biology* **46**, 479–506 (1984).

M. M. Elkind and H. Sutton, Radiation response of mammalian cells grown in culture. *Radiation Research* **13**, 556–593 (1960).

R.E. Franklin and R. G. Gosling, Molecular structure of deoxypentose nucleic acids. *Nature* **171**, 738–741 (1953).

S. Furberg, On the structure of nucleic acids. *Acta Chem. Scand.* **6**, 634–640 (1952).

E. I. Hart and J. W. Boag, Absorption spectrum of the hydrated electron in water and in aqueous solutions. *J. Am. Chem. Soc.* **84**, 4090–4095 (1962).

T. Henriksen, Hydrogen atoms and solvated electrons in irradiated aqueous solutions at 77 K. *Radiation Research* **23**, 63–77 (1964).

A. D. Hershey and M. Chase, Independent functions of viral proteins and nucleic acids in growth of bacteriophage. *J. General Physiology* **36**, 39–56 (1952).

R. Livingston, H. Zeldes and E. H. Taylor, Paramagnetic resonance studies of atomic hydrogen produced by ionizing radiation. *Dis. Faraday Soc.* **19**, 166 (1955).

H. J. Muller, Artificial transmutation of the gene. *Science* **66**, 84–87 (1927).

G. Olivieri, J. Bodycote and S. Wolf, Adaptive response of human lymphocytes to low concentrations of radioactive thymidine. *Science* **223**, 594 (1984).

H. Planel, J. P. Soleilhvoup, R. Tixador, G. Richoilley, A. Conter, F. Croute, C. Caratero and Y. Gaubin, Influence on cell proliferation of background radiation or exposure to very low, chronic gamma radiation. *Health Physics* **52**, 571–578 (1987).

R. L. Platzman, Subexcitation electrons. *Radiation Research* **2**, 1–7 (1955).

J. L. Redpath and R. J. Antoniono, Induction of an adaptive response against spontaneous neoplastic transformation *in vitro* by low-dose gamma radiation. *Radiation Research* **149**, 517 (1998).

E. Sagstuen, H. Theisen and T. Henriksen, Dosemetry by ESR spectroscopy following a radiation accident. *Health Physics* **45**, 961–968 (1983).

A. C. Upton, The biological effects of low-level ionizing radiation. *Scientific American* **246**, 29 (1982).

J. D. Watson and F. H. Crick, Molecular structure of nucleic acids. *Nature* **171**, 737–738 (1953).

Index

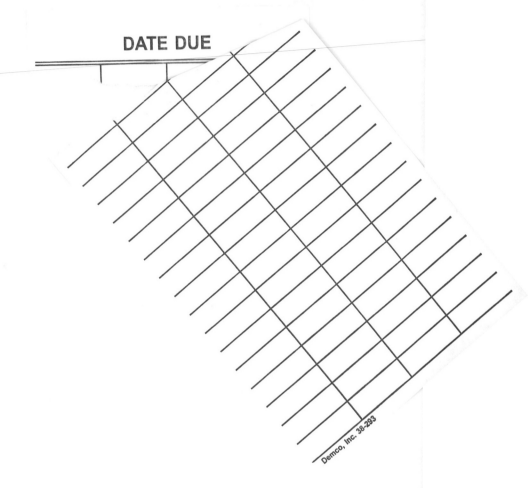

DATE DUE

Demco, Inc. 38-293